T0328271

Bio-Engineering Approaches to Cancer Diagnosis and Treatment

Bio-Engineering Approaches to Cancer Diagnosis and Treatment

Azadeh Shahidian

Majid Ghassemi

Javad Mohammadi

Mohadeseh Hashemi

ACADEMIC PRESS

An imprint of Elsevier

Academic Press is an imprint of Elsevier
125 London Wall, London EC2Y 5AS, United Kingdom
525 B Street, Suite 1650, San Diego, CA 92101, United States
50 Hampshire Street, 5th Floor, Cambridge, MA 02139, United States
The Boulevard, Langford Lane, Kidlington, Oxford OX5 1GB, United Kingdom

Notices
Knowledge and best practice in this field are constantly changing. As new research and experience
broaden our understanding, changes in research methods, professional practices, or medical treatment
may become necessary.

Practitioners and researchers must always rely on their own experience and knowledge in evaluating and
using any information, methods, compounds, or experiments described herein. In using such information or methods they should be mindful of their own safety and the safety of others, including parties for
whom they have a professional responsibility.

To the fullest extent of the law, neither the Publisher nor the authors, contributors, or editors, assume
any liability for any injury and/or damage to persons or property as a matter of products liability, neg-
ligence or otherwise, or from any use or operation of any methods, products, instructions, or ideas
contained in the material herein.

Library of Congress Cataloging-in-Publication Data
A catalog record for this book is available from the Library of Congress

British Library Cataloguing-in-Publication Data
A catalogue record for this book is available from the British Library

ISBN: 978-0-12-817809-6

For information on all Academic Press publications
visit our website at https://www.elsevier.com/books-and-journals

Publisher: Mara Conner
Acquisitions Editor: Fionna Geraghty
Editorial Project Manager: Isabella C. Silva
Production Project Manager: Prasanna Kalyanaraman
Designer: Christian J. Bilbow

Typeset by Thomson Digital

Working together
to grow libraries in
developing countries

www.elsevier.com • www.bookaid.org

To my mother Mansoureh, my father Hossein, my husband Hossein,
my daughter Fatemeh, and my sons Ali and Mahdi

Azadeh

To my wife Monir, my daughter Fatimah and my son Alireza

Majid

To my mother Parizad, my father Nosrat, my wife Maryam,
my daughters Hosna and Hanieh, and my son Amir Mohammad

Javad

To my father Shokrollah, my mother Zohreh, my sister Motahareh,
my brother Ali, and my friend Shahla

Mohadeseh

Contents

Acknowledgments

We, the authors, would like to express our greatest respect to our parents *without them none of this would be possible*. The authors would also like to extend their highest gratitude to their immediate family for providing endless support and encouragement during the entire endeavor. Our gratitude appreciates the assistance from our master and PhD students at the Nano and Fuel Cell Laboratory in the K.N. Toosi University of Technology. Our special thanks to Dr. Roozbeh Ayani, Dr. Seyed Mohammad Ali Nemati, Mrs. Sanam Tahouneh, Mrs. Farnaz Jazini Dorcheh, and Mr. Mehrshad Rezadoost Dezfuli.

Dr. Javad Mohammadi would like to thank Mrs. Shirin Nouraein and Mrs. Negin Vahedi for their major contribution to the main part of Chapter 3 and a part of Chapters 4 and 5. He also would like to thank Mrs. Bahareh Sadri for her major contribution to the main part of Chapter 4 and a portion of Chapter 5. Dr. Mohammadi also appreciates Taravat Khodaei for her major contribution to the two subchapters of Chapter 4. He specifically expresses his outmost appreciation to Dr. Hodjatallah Rabbani for his scientific consultation and Mr. and Mrs. Shahin and Samin Kazem Zadeh for their efforts in editing of above chapters.

We also appreciate the Elsevier staff, Maria Convey, Fiona Geraghty, Kavitha Balasundaram, Isabella Conti Silva, and Prasanna Kalyanaraman for their support and assistance throughout the entire process. We also appreciate the referee's time and useful suggestion.

Last and not least: I beg forgiveness of all those who have been with us over the course of the book and whose names we have failed to mention.

Introduction

Chapter outline

1.1 Cancer immunology

Cancer is a lethal disease, which can affect all human tissues. It is a life-threatening health problem that is linked to several genetic and environmental risk factors related to immune function. These diseases produce immunosuppressive factors to disrupt cell division, leading to uncontrolled proliferation of cancer cells. Most importantly, tumor cells have learned to evade immune attack by presenting similar antigens to normal cells and expressing very low levels of antigens. cancer is the second cause of death (25%) in the United States [1]. The main kinds of cancers in adults include breast, lung, prostate, and colorectal malignancies [2]. Lymphoma was the most common invasive cancer (20%), after that invasive skin cancer (15%), endocrine system cancer (11%) and male genital system cancer (11%) [2]. During 2004–08, overall cancer incidence has reduced in men by 0.6% per year. However, this rate is stable due to the high prevalence of breast cancer among women [3]. Based on recent reports, during the past decade, the cancer incidence rate since 2006–2015 in women was stable and in men declined by approximately 2% per year, whiles the rate of death due to cancer (2007–16) declined annually by 1.4% and 1.8%, respectively. Also, the overall cancer mortality rate has steadily decreased by 27% from 1991 to 2016. While the ethnic gap in cancer-related mortality is gradually narrowing, socio-economic inequities are expanding and the most significant gaps for the most preventable cancers. For instance, during 2012–16, compared to the richest cities, the death rate in the poorest cities was twice as high for cervical cancer and 40% higher for lung and liver malignancies [4].

Many predisposing factors have been identified. The incidence of cancer is found to be significantly associated with the age of 10–60 years. In addition, the male gender is at higher risk of developing cancer than women [2]. Race is another important factor for cancer progression. Highest incidence has been reported before 40 years of age, among non-Hispanic whites and, after 40 years of age, amongst African-Americans/blacks [5]. Cancer can be a dangerous health problem, particularly when the tumor has metastasized to other organs. Four cancers: lung, bronchial, breast and colorectal in women and lung, bronchial, prostate and colorectal in men account for approximately 50% of cancer-related deaths.

Cancer immunology has long been studied. However, the molecular and cellular basis of tumor immunity is not fully understood. Advances in understanding the basics of the immune system and advances in the treatment of infectious diseases have had a major effect on the development of immunotherapy. The role of T cell responses in tissue allograft rejection has been known since the 1950s. Since then, it has been recognized that tumors occur in association with T-cell dysfunction, indicating the importance of the immune system in the growth and development of cancer [6].

The identification of tumor-associated antigens (TAA), understanding of effector T cell responses, and the role of regulatory and suppressor T cells have determined the use of the immune system to treat cancer. Tumors refer to an enormous spectrum of diseases that originate from uncontrolled cellular growth. Tumor is divided into benign tumors and malignant tumors [7].

Cancers are further classified by their cell type and tissue of origin.

1.2   Cancer stem cell

Cancer stem cells (CSCs) have been described as the origin of several types of human cancer. However, other reports have indicated that CSCs are derived from stem cells, tissue commitment progenitor cells, or tissue cells [8–14].

Stem cells are recognized by two main properties: self-renewal and differentiation.

In CSCs, there is an unbalance between self-renewal and differentiation [15]. The main step of alteration of natural stem cell to CSC is self-renewal step. Imbalance between stimulating and inhibitory signaling in self-renewal and proliferation among stem cells leads to malignancy [15].

CSCs are a small population of cancer cells. Only a small part of tumor with CSCs phenotype, stemness, have self-renewal character can produce tumor and treatment of tumor cannot destroy CSCs, but can kill other tumor cells [16,17].

The uncontrolled cell division can originate from several factors, such as viral infections, chemical agents and mutations that lead to evasion of cells from the checkpoints which control cell division [6]. It has been reported that some tumors are initiated by oncogenic viruses—RNA and DNA viruses—that induce malignant transformation and prophylactic immunization against some viruses such as EBV, HPV, and HBV may be a rational strategy for prevention of malignancy [18].

The disrupted immune response can induce tumor growth and prevent effective antitumor suppression, possibly through a process of "sneaking through" that allows better growth of micro-tumors rather than large tumors [19].

1.3   Cancer and immune system impairment

Based on investigations, tumors may produce some immunosuppressive factors, such as alpha-fetoprotein, interleukin-10 (IL-10) and transforming growth factor-β (TGF-β), which suppress immune responses against tumors and cancer cells themselves can, therefore, lead to immunosuppression. Hence, scientists have used neutralizing antibodies against these immunosuppressive factors [6]. Interestingly, patients with primary immunodeficiency diseases have higher risk to develop cancer.

CSCs are different from normal stem cell and can be identified by the immune system. The immune system also produces a set of proteins such as complements with effects against cancer and other immune or inflammatory responses. These proteins released by T cells and are classified as "cytokines." Cytokines include proteins such as tumor-necrosis factors, interferons, interleukins, and growth factors. Based on investigations, vaccination accompany with the complements can cause tumor lysis. While incompletely defined, several soluble and cellular mediators of tumor rejection have been described, including complement factors, active macrophages, T cells, and NK cells. Though T cells need antigen specificity, the molecular and cellular mechanisms of innate immunity can detect the tumor phenotype in the lack of antigen specificity [20]. Because most TAA are self-proteins, the immune response cannot detect and destroy them and in patients' immune tolerance can be created to TAAs. In addition, the cells of the immune system may not influence tumor microenvironment, resulting in slower immune-mediated tumor eradication [19].

Reports have also indicated that genetic elements are as important as environmental carcinogens. Investigations on retrovirus infection between different breeds of animals have indicated that a unique carcinogen resulted in different consequences among different breeds, demonstrating the effect of genetic background in the induction of cancer. Environmental factors may also regulate immune responses [21].

1.4   Cancer therapy and immunotherapy

As previously have been mentioned, tumor cells can express similar antigens to normal cells, with lower quantity allowing tumor cells to evade immune system attack by inducing of tolerance. Immunotherapy may damage to the normal tissues and organs, such as process by the development of autoimmunity induced by anticytotoxic T lymphocyte-associated antigen 4 (anti-CTLA-4) or anti-PD1 monoclonal antibody (mAb) treatment [22]. There are several suggested mechanisms for the escape of cancer cells from host immunity. These ways include decreased expression of immunogenic TAAs, beta2-microglobulin, MHC class I molecules, and costimulatory molecules which have influence on activation of T cells. Other mechanism is the expression of very low levels of antigens, unable to induce immune response. Cancer cells may proliferate rapidly and induce a rapid immune response. Another mechanism for evade of cancer cells from surveillance of immune system include of tumor signaling by CTLs, secretion immunosuppressive molecules dampening tumor-reactive effector T cells and the induction of regulatory or suppressor cells [23]. So far, direct evidence about tumor-regulated immunity is derived from experimental studies in animal models. These animal models support antitumor immunity by vaccination using inactive cancer cells or by resection of primary tumors. Furthermore, antitumor immunity can be induced by administration of tumor-reactive T lymphocytes [24].

Currently, new molecular immunological diagnostic methods for tumor diagnosis are proposed.

In vivo detection of micro tumors is available by using of monoclonal antibodies marked with radioisotopes.

In vitro detection of the undifferentiated tumor cell using monoclonal antibodies is other application of these molecules. Another application of immunodiagnostics methods is to detection of metastatic cells [25].

There are four general therapeutic approaches in cancer therapy:

1. Chemotherapy, the treatment of disease by the use of chemical substances, especially the treatment of cancer by cytotoxic and other drugs that interfere with cell division and DNA synthesis.
2. Hormonal therapy, is a cancer treatment that slows or stops the growth of cancer that uses hormones to grow and interfere with growth signaling via tumor cell hormone receptors.
3. Targeted therapy, a type of treatment that using drugs or other substances to recognize and attack specific types of tumor cells with low damage to normal cells. Some targeted therapies block the action of certain enzymes, proteins, or other molecules involved in the growth and metastasis of tumor cells.
4. Immunotherapy, immunotherapy is a type of biological therapy and cancer treatment that helps immune system fight against cancer. This method uses substances made from living organisms to treat tumors [26].

Cancer immunotherapy is a novel therapeutic method for cancer treatment, with promising results. This approach has improved the cancer-related immune response by either increasing mechanisms of the immune system that induces an effective immune response or by suppressing mechanisms that inhibit the immune response. Currently, two approaches are used for immunotherapy including allogeneic bone marrow transplantation and monoclonal antibodies (mAbs) targeting tumor cells or T cell checkpoints [27]. Firstly, cancer vaccines have been considered for prevention and treatment of several cancers [27]. Based on reports, more than 15% of human cancers are caused by viruses and cancer vaccine may, thus, be useful for these kinds of tumors [28]. A genetically engineered oncolytic herpes simplex type 1, which secrete granulocyte-macrophage colony-stimulating factor (GM-CSF), has been used for cancer therapy. This immunotherapeutic approach has had beneficial effects in the eradication of head and neck squamous cell carcinoma and melanoma [29]. One more approach is using mAbs which block T cell checkpoints role to regulate T cell responses. CTLA4 is a significant inhibitory receptor that has regulatory effect on T cell function and plays an important role in the priming phase of the immune response. This molecule is expressed on T helper and T cytotoxic cells, as well as on FOXP3+ regulatory T cells [30]. Administration of mAbs by targeting human CTLA4 has improved survival in patients with metastatic melanoma [31]. Another mAbs blocking other T cell checkpoints, such as the programmed cell death protein 1 (PDCD1/PD1), V-set domain-containing T cell function inhibitor 1 (B7x), programmed cell death ligand 1 (PDL1/CD274), CD276 (B7H3) antigen, and B, T lymphocyte attenuator, have also been used in clinical trials. Furthermore, some investigations have indicated a significant therapeutic effect in some kinds of tumors, including ovarian cancer, nonsmall cell lung carcinoma, melanoma and renal cell carcinoma [32]. PDL1 expression by tumor cells has been reported to be associated with poor clinical outcomes and may be accompanied with clinical response to anti-PD1 and anti-PDL1 treatment. On the other hand, PDL1 inactivate tumor-infiltrating cells and regulatory T cells (T_{reg} cells) have an immunosuppressive role in the tumor microenvironment [33]. Additionally, the combination of these agents with anti-CTLA4 and other immunotherapy approaches has led to promising results. The combination of cancer vaccines with factors targeting the IL-12 receptor has resulted in contradictory results. This may be because of the upregulation of IL-12 receptor by both activated T effector cells and T_{reg} cells [34]. Therefore, new strategies focused on more specific targeting of T_{reg} cells which decrease their immunosuppressive effects are needed. Adoptive T-cell therapy has been qualified as an efficient treatment for cancer immunotherapy in early phase clinical trials. In this approach, a large quantity of tumor-specific T cells derived from peripheral blood or preferably from the tumor niche is transferred to patients with established tumors [35]. Based on reports, a combination of chemotherapy and immunotherapy has beneficial effects and in future chemoimmunotherapy is an attractive method [36]. These combination methods could be Toll-like receptor (TLR) signaling pathway agonists/antagonists, cytokines, mAbs targeting T cell checkpoints and cancer vaccines [37]. In addition, radiation and radiofrequency ablation can be other combination therapy with immunotherapy [38]. Although immunotherapy and its combination with other therapies such as radio immunotherapy may be useful for tumor treatment, there are many limitations to consider.

Patient definition as an optimal target, optimal biological dose, and timetable, need for better trial designs including appropriate clinical endpoints, and identification and validation of predictive biomarkers are just a few to mention [27].

By proliferation of cancer cells, the formation of the orchestrated enzyme required for proper metabolism of its various components may become unbalanced and produce products that are not observed in normal dividing cells [39]. Recently, it has been reported that these biochemical "switches" lead to uncontrolled multiplication of cancer cells as seen in a type of leukemia. Hence, it has been suggested that targeting tumor switches can lead to the treatment of cancers [24]. However, it is unclear how this approach can be used to optimize tumor immunotherapy.

1.5   Immunoassay diagnosis technics in cancer

Cancer is a disease that can be cured if it is rapidly diagnosed. There are many diagnostic technics available today, however, the survival rate of late-stage cancers remains low due to late diagnosis [40]. What can be known as ideally is diagnosing and treating cancer at its early stages. Hence, this is the main purpose of much of the technics and research happening until today. From the 20th century, it was understood that cancer is a genetic disease which means that it happens based on genomic changes on some genes especially those which control the cell function in growing and divide which underpins the core element of today's diagnostic methods [41]. The Human Genome Project, has completed in 2003, marked a dramatic shift in the understanding of cancer and other diseases. After 13 years, researchers mapped the entire human genetic code, discovering that every human cell is packed with an estimated 20,000–30,000 genes. Researchers have used the discoveries to link dozens of diseases, such as Alzheimer's disease and inherited colon cancer, to specific genes.

Genes contain instructions that play an important role in protein production. In cancer study, it was discovered that these genome changes cause the changes in protein production. For instance, in some cancers, it causes more growth protein production or misshapen protein that all of them lead to evade cells from a normal situation. With regard to cancer, researchers have discovered that two genes, BRCA1 and BRCA2, are associated with breast and ovarian cancers. They have understood that the alternation in these genes causes overproduction of protein HER2 [42]. With this knowledge, they have investigated drug for treatment of breast cancer. The advances in genomic technologies have numerously facilitated the understanding of cancer progression and metastasis, and the discovery of novel biomarkers today is known as one of the main technics in cancer diagnosis. Genomic testing in the diagnosis of cancer has two main approaches. The first technic looks at very small numbers of genes, called single/limited gene panel testing. The second technic analyses a larger number of genes called multigene panel testing. These technics could be used as a diagnostic method in the early stage. It is known that inherited genetic abnormalities can lead to the cancer which detects mutations passed down from one generation to the next. Genomic based technics were large progress in cancer diagnosis, however, they have certain disadvantages. Genomic-based alterations are difficult to correlate with a particular disease feature, and it has been difficult to specify which proteins interact. This means that they provide limited information about the disease or heterogeneous tumors present in patients [40]. For example, it can be seen that two patients can display identical symptoms and imaging results of a certain cancer at the time of diagnosis; however, at chemotherapy show drastically different outcomes. This is because beside the genomic changes, the molecular sorting of individual tumors is different and therefore needs different targeting chemotherapy in order to have a positive outcome. Today we know that cancer is generally activated by the acquisition of somatic DNA lesions. Cancer is a combination of both genetic and epigenetic alterations which cause instability in the genome and implicated genes, such as oncogenes, tumor suppressor genes, apoptotic genes, and DNA repair genes [43]. Based on recent studies, tumor-associated somatic mutations or copy number variations might be responsible for human colon cancer and there is a poor correlation between expressed proteome, protein profile and DNA mutation in genome of the disease. The sequencing of individual cancer genomes has underscored the remarkable complexity and heterogeneity in the same cancer subtypes and histopathological phenotypes, therefore it cannot be detected easily with genomic technics. It is known that, alterations in somatic mutational sites of the oncogenome are translated into contrarily regulated oncoproteome initiating the cancer occurrence [40].

Thus, combination of proteomic and genomic technologies is required for early-stage diagnosis. Biomarkers are required tools for cancer diagnosis and monitoring. It is also known that the physiological status of the cell has an important role in cancer diagnosis in patients. Thus biomarkers could be the hallmark at both diagnosis and treatment process. Gene mutation in gene transcription and translation, and a variety of posttranslational protein modifications (PTMs) potentially contribute to specific "biochemical footprints" of disease, which highlights the importance of biomarkers in cancer diagnosis. In human genome-based detection of cancer using advanced genomic technologies such as Next Gen Sequencing, digital PCR, circulating free DNA technology, has not shown a true detection. Therefore, proteomic methods and the proteogenomic technics are promising tools. Proteomic technics provide a database of PTMs and Somatic variants that became a good complementary for histopathology.

Besides the proteome and genome technics, antibody-based immunosensor is another technic that has been investigated for the detection of biomarkers in cancer diagnosis. Imaging also in recent years with the aim of biomarker detection for tumor diagnosis has been more in attention than before.

Reliability, sensitivity and specificity of noninvasive methods for detecting CSCs utilizing currently available protein biomarkers and their PTMs are important parameters.

During studies that have been conducted around cancer diagnosis, it is important to detect genes and biomarkers which are related to cancer disease. Therefore, nowadays there is a strong focus on high sensitive assay to detect these genes and biomarkers and also usage of multiple approaches to increase their sensitivity and their ability to detect more related biomarkers. Researcher has also shown that PTM plays an important role in diagnosis and treatment of cancer. Here we demonstrate the updated technics with the aim of high sensitivity and being less invasive for patients.

1.6 Bioengineering assisted cancer imaging using nonbiological components

Finding the optimal methods of cancer imaging is always received a great attention not only for diagnostic purpose but also for treatment. The optimal biomedical imaging needs to have important criterion including less destruction of tissue and invasiveness, accurate real-time monitoring, feasibility to detect over wide ranges of tumors, and targeting ability. Size scales are varied from molecular to whole organism level.

The role of biological component in cancer diagnostic already discussed in Chapter 2. Herein, the brief review of nonbiological systems in cancer imaging will be reviewed. X-ray imaging technic is based on the ability of the targeted tissue to attenuate the irradiated X-ray leading to the formation of the image with different opacity. Another imaging technic is magnetic resonance imaging (MRI) which is based on the relaxation time of water in a magnetic field. In general, each compartment has been surround by water and has a specific relaxation time. Different relaxation time brings the opportunity to detect the compartments.

Ultrasound is longitudinal acoustic energy with the frequency above 20 kHz. Irradiation of ultrasound cause particle displacement along with the wave propagation. Ultrasound imaging is based on the scattering, reflection, and shifts of acoustic waves. Another application of ultrasound is thermal therapy which will be discussed in the next topic of this chapter [44].

Another imaging technic is based on nonionizing electromagnetic radiation contain low photon energy from 1 to 3×10^{15} Hz. One of the main differences between the interaction of ionizing and nonionizing radiation with an atom is that ionizing radiation cause removing of electron while nonionizing radiation cause electron excitation and heat production.

The physical properties of imaging systems including resolution and sensitivity are different. The most sensitive imaging systems are nuclear medicine and positron-emission tomography (PET). In PET, image is constructed based on the detection of γ photons released from radionuclides during the annihilation of positron with electron. A positron annihilates with electrons leading the release of two photons moving in the opposite directions. The photons will be detected using scintillation crystals.

To increase the accuracy of detecting systems, hybrid imaging systems have been introduced. For example, PET/CT which is the combination of the spatial resolution of CT and metabolic sensitivity of PET.

The main obstacles in optical imaging are mainly related to the absorption and scattering of light by untargeted tissues and fluids. To tackle the problem, smart nanoparticle targeted to desired tumor is under investigation [45].

1.7   Bioengineering assisted cancer treatment using nonbiological components

Applying heat to treat cancer has been started since 1700 BC by using a fired-drill to treat breast cancer. Hyperthermia is based on the increasing the temperature of the desired tissue till 41–47°C to selectively destruct tumor cells. The main idea behind this technic is related to the low level of heat tolerance in tumor cells as compared with normal cells. Cell damage during hyperthermia may related to the protein denaturation. There are several methods to induce heat to desired site such as ultrasound waves, microwaves, radiofrequency and laser.

"Laser" is an acronym for the light amplification by the stimulated emission of radiation which emits photons in a coherent beam. Using laser to apply heat to tumorogenic region was firstly reported in 1965.

Laser may assist cancer treatment by several approaches including laser ablation, surgery, photodynamic therapy (PDT), photothermal therapy (PTT), and triggering an anticancer drug delivery. Laser ablation or photoablation is removing tumor cells using laser beam with high intensity, usually pulse laser. Laser surgery performs specific functions during surgery. PTT is a kind of cancer treatment using photothermal agents to selectively destroy cancer cells. Photothermal agents convert optical energy into heat. PDT is another kind of treatment using photosensitizer to destroy cancer cells. When light exposes to photosensitizers, it reacts with tissue oxygen and produces toxic singlet oxygen.

Several kinds of laser have been used for cancer treatment included carbon dioxide (CO_2) lasers, neodymium:yttrium-aluminum-garnet (Nd:YAG) lasers, laser-induced interstitial thermotherapy (LITT), and argon lasers [46].

1.8   Principles of heat and fluid flow

1.8.1   Fluid mechanics

Fluid mechanics are the branch of physics that studies the mechanics of fluids (liquids, gases, and plasmas) and the forces on them. Fluid mechanics can be divided into fluid statics or the study of fluids at rest; and fluid dynamics or the study of the effect of forces on fluid motion. Fluid mechanics have a wide range of applications, including mechanical engineering, chemical engineering, geophysics, astrophysics, and biology. Fluid mechanics, especially fluid dynamics, are an active field of research with many problems that are partly or wholly unsolved. Commercial code based on numerical methods is used to solve the problems of fluid mechanics. The principles of these methods are developed by CFD. A modern discipline, called computational fluid dynamics (CFD), is devoted to this approach to solve the aforementioned problems.

The distinguished characteristic of fluids compare to solids is related to the amount of deformation rate. Fluids show a continuous deformation when they get exposed to a shear force. However, different fluids reveal different rates of deformation when a specified shear force acts on them. According to Newton's law of viscosity, for common fluids such as water, oil, and gasoline the ratio of shear stress existed in fluid layers to the velocity gradient is equal to a property of fluid named dynamic viscosity (for more details, readers are referred to Ref. [47]):

$$\frac{\tau}{du/dy} = \mu \tag{1.1}$$

τ is shear stress, μ is dynamic viscosity, and du/dy is velocity gradient.

Such fluids, that shear stress linearly varies with velocity gradient, are categorized as Newtonian fluids. On the other hand, for non-Newtonian fluids, shear stress does not vary linearly with velocity gradient.

In order to have a comprehensive understanding of fluid mechanics problems, differential analysis of fluid flow is necessary. Therefore, fluid mechanics governing equations, conservation of mass, and momentum must be analyzed.

Conservation of mass: (also called continuity equation) in vector notation is [47]:

$$\frac{\partial \rho}{\partial t} + \nabla \cdot (\rho V) = 0 \tag{1.2}$$

ρ is the fluid density, V is velocity vector, and ∇ is gradient vector. First and second terms in Eq. (1.2) denote the rate of change in fluid density and net rate of mass outflow per unit volume, respectively. Density for incompressible fluids is constant. Therefore Eq. (1.2) becomes:

$$\nabla \cdot V = 0 \tag{1.3}$$

Conservation of momentum: based on Newton second law the conservation of momentum is [47]:

$$\rho \frac{\partial V}{\partial t} + (V \cdot \nabla)V + g = \nabla \cdot [-PI + \tau] \tag{1.4}$$

In which P is pressure and τ is stress tensor and **I** is identity matrix. The first and the second terms on the left-hand side of Eq. (1.4) denotes local and convective acceleration and the inertia force per unit volume, respectively. Based on Stokes theorem, shear stress for a Newtonian fluid is expressed as [48]:

$$\tau = 2\mu S - \frac{2}{3}\mu(V \cdot \nabla)I \tag{1.5}$$

S is the strain-rate tensor and is [1]:

$$S = \frac{1}{2}\left(\nabla V + (\nabla V)^T\right) \tag{1.6}$$

By plugging Eqs. (1.5), (1.6) into the momentum equation (Eq. 1.4), it becomes [47]:

$$\nabla \cdot \left[-PI + \mu\left(\nabla V + (\nabla V)^T\right) - \frac{2}{3}\mu(V \cdot \nabla)I\right] + g = \rho \frac{\partial V}{\partial t} + \rho(V \cdot \nabla)V \tag{1.7}$$

In which pressure (P) and velocity components (u, v, and w) are dependent variables. Eq. (1.7) is the general form of momentum equation and is used for both compressible and incompressible fluids. For incompressible fluids (ρ = const) third term from the left-hand side of Eq. (1.7) vanishes and it reduces to [47]:

$$\nabla \cdot \left[-pI + \mu \left(\nabla V + (\nabla V)^T \right) \right] + g = \rho \frac{\partial V}{\partial t} + \rho (V \cdot \nabla) V \tag{1.8}$$

In general, to solve Navier-Stokes equation initial and boundary conditions must be available. The initial boundary condition is the condition of the system at time zero. Typical boundary conditions in fluid dynamic problems are solid boundary conditions, inlet and outlet boundary conditions, and symmetry boundary conditions and are specifically defined for each problem [47].

1.8.1.1 Euler and Bernoulli equations

In the case of inviscid fluids ($\mu = 0$) the equation of motion (Eq. 1.4) is simplified to [49]:

$$\nabla P + g = \rho \frac{\partial V}{\partial t} + \rho (V \cdot \nabla) V \tag{1.9}$$

Eq. (1.9) is commonly referred to Euler's equation of motion. Under steady-state condition, the first term of the right-hand side of the equation is to be zero and the Euler's equation is simplified as follows [49]:

$$\nabla P + g = \rho (V \cdot \nabla) V \tag{1.10}$$

By integrating this equation along some arbitrary streamline and also assuming incompressible fluid we obtain [49]:

$$\frac{P}{\gamma} + \frac{V^2}{2g} + z = const \tag{1.11}$$

where γ is the specific weight of fluid and z is the elevation of the point above a reference plane. Eq. (1.11) is called Bernoulli equation and is valid for inviscid, steady, incompressible flow along a streamline or in the case of irrotational flow along with any two arbitrary points. In another word, Bernoulli indicates that the pressure stays constant during the flow when the tube cross-section and height do not change.

1.8.1.2 Nondimensional parameters

A fundamental nondimensional parameter in analyses of fluid flow is the Reynolds number (Re) [49]:

$$Re = \frac{\rho UL}{\mu} \tag{1.12}$$

U is velocity scale and L denotes a representative length. The Reynolds number represents the ratio between inertial and viscous forces. At low Reynolds numbers, viscous forces dominate and tend to damp out all disturbances, which leads to laminar flow. At high Reynolds numbers, the disturbances appear and at high enough Reynolds number the flow field eventually ends up in a chaotic state called turbulence.

Another important nondimensional number in biofluid mechanic is the Womersley number (α), relating the pulsatile flow frequency in with viscous effects [50].

$$\alpha = L\left(\frac{\omega}{\nu}\right)^{1/2} = \left(2\pi\,\mathrm{Re}\,St\right)^{1/2} \tag{1.13}$$

The Strouhal number (St), is a dimensionless number describing the mechanism of oscillating flow, and is given by [50]:

$$St = \frac{fL}{U} \tag{1.14}$$

where f is the frequency of vortex shedding, L and U are the characteristic lengths and the flow velocity. The Euler number (Eu) is defined in hydrodynamic and described the stream pressure versus inertia forces.

$$Eu = \frac{\Delta P}{\rho V^2} \tag{1.15}$$

Also the Froude number (Fr) is defined as the ratio of the flow inertia to the external field (such as gravity). This number is based on the speed-length ratio as follow:

$$Fr = \frac{U}{\sqrt{gl}} \tag{1.16}$$

1.8.2 Heat transfer

In physics, heat is defined as the transfer of thermal energy across a well-defined boundary around a thermodynamic system. Heat transfer is a process function (or path function). It means that the amount of heat transferred that changes the state of a system depends on how that process occurs, not only the net difference between the initial and final states of the process. The rate of heat transfer is dependent on the temperatures of the systems and the properties of the intervening medium through which the heat is transferred. The rate of heat transfer also depends on the properties of the intervening medium through which the heat is transferred. In engineering contexts, the term heat is taken as synonymous to thermal energy. This usage has its origin in the historical interpretation of heat as a fluid (caloric) that can be transferred by various causes.

The transport equations for thermal energy (Fourier's law), mechanical momentum (Newton's law for fluids), and mass transfer (Fick's laws of diffusion) are similar, and analogies among these three transport processes have been developed to facilitate prediction of conversion from anyone to the others. The fundamental modes of heat transfer are advection, conduction or diffusion, convection, and radiation. Types of phase transition occurring in the three fundamental states of matter include deposition, freezing and solid-to-solid transformation in solid, boiling/evaporation, recombination/deionization, and sublimation in gas and condensation and melting/fusion in liquid. Heat transfer through living biological tissue involves heat conduction in solid tissue matrix and blood vessels, blood perfusion (convective heat transfer between tissue and blood), cooling of human body by radiation as well as metabolic heat generation. Heat transfer processes are classified into three types: conduction, convection, and radiation.

1.8.2.1 Conduction heat transfer

Conduction heat transfer is the transfer of heat through matter (i.e., solids, liquids, or gases) without bulk motion of the matter. In another ward, conduction is the transfer of energy from the more energetic to less energetic particles of a substance due to interaction between the particles. Conduction heat transfer in gases and liquids is due to the collisions and diffusion of the molecules during their random motion. On the other hand, heat transfer in solids is due to the combination of lattice vibrations of the molecules and the energy transport by free electrons. For example, heat conduction can occur through wall of a vein in human body. The inside surface, which is exposed to blood, is at a higher temperature than the outside surface.

To examine conduction heat transfer, it is necessary to relate the heat transfer to mechanical, thermal, or geometrical properties. Consider steady-state heat transfer through the wall of an aorta with thickness Δx where the wall inside the aorta is at higher temperature (T_h) compare to the outside wall (T_c). Heat transfer $\dot{Q}$(W), is in direction of x and perpendicular to plane of temperature difference. Heat transfer is function of aorta wall higher and lower temperature, the aorta geometry and properties and is given by [51]:

$$\dot{Q} \propto \frac{(A)(\Delta T)}{\Delta x} \tag{1.17}$$

or

$$\dot{Q} = kA\frac{(T_h - T_c)}{\Delta x} = -kA\frac{(T_c - T_h)}{\Delta x} = -kA\frac{\Delta T}{\Delta x} \tag{1.18}$$

In Eq. (1.18), thermal conductivity $\left(k, \text{W/m K}\right)$ is transport property. Parameter A is the cross-sectional area (m^2) of the aorta and Δx is the aorta wall thickness (m). In the limiting case of $\Delta x \to 0$ Eq. (1.18) reduces to Fourier's law of conduction:

$$\dot{Q} = -kA\frac{dT}{dx} \tag{1.19}$$

where $\frac{dT}{dx}$ is the temperature gradient and must be negative based on second law of thermodynamics.

A more useful quantity to work with is heat flux, q'' (W/m^2), the heat transfer per unit area:

$$q'' = \frac{\dot{Q}}{A} \tag{1.20}$$

Separating the variables in Eq. (1.19), integrating from $x = 0$ and rearranging gives:

$$\dot{Q} = kA\frac{\Delta T}{L} \tag{1.21}$$

Thermal resistance circuits: For steady one-dimensional flow with no generation heat conduction equation, Eq. (1.21) can be rearranged as:

$$\dot{Q} = \frac{\Delta T}{L/KA} = \frac{\Delta T}{R_{cond}} \tag{1.22}$$

Conduction thermal resistance (R_{cond}) is represented by:

$$R_{cond} = \frac{L}{KA} \tag{1.23}$$

It is obvious that the thermal resistance R_{cond} increases as wall thickness (L) increases, area (A) and K decreases. The concept of a thermal resistance circuit can be used for problems such as composite wall thickness.

The heat transfer rate for composite wall is given by:

$$\dot{Q} = \frac{\Delta T}{\sum R_{cond}} = \frac{\Delta T}{R_1 + R_2} \tag{1.24}$$

1.8.2.2 Convective heat transfer

Convection heat transfer is due to the moving fluid. The fluid can be a gas or a liquid; both have applications in bio and nano heat transfer. Convection is the energy transfer between two mediums; typically, a surface and fluid that moves over the surface. In convective heat transfer heat is transferred by diffusion (conduction) and by bulk fluid motion (advection). An example of convection heat transfer is the flow of blood inside the human vessels or air and water flow over the human skin. In convective heat transfer it is important to examine the fluid motion near the surface. Close to wall there exists a thin layer called "boundary layer" where fluid experience velocity and temperature differences. Boundary layer thickness depends on flow Reynolds number, structure of the wall surface, pressure gradient and Mach number. Outside this layer, temperature and velocity are uniform and identical to free stream temperature and velocity.

The rate of convection heat transfer ($\dot{Q}$) from/to the surface is given by Newton's Law of Cooling as [51]:

$$\dot{Q} = hA(T_w - T_\infty) \tag{1.25}$$

The quantity h (W/m^2 K) is called convective heat transfer coefficient and T_w and T_∞ are surface and fluid temperature, respectively. For many situations of practical interest, the quantity h is known mainly through experiments. Integrating

Eq. (1.25) over the entire surface leads to average convection coefficient $\left(\bar{h}\right)$ for the entire surface:

$$\bar{h} = \frac{1}{L}\int_{A_s} h\, dA_s \tag{1.26}$$

A thermal resistance is also defined for convection heat transfer from Eq. (1.27):

$$\dot{Q} = \frac{\Delta T}{1/hA} = \frac{\Delta T}{R_{conv}} \tag{1.27}$$

The convective thermal resistance (R_{conv}) is then:

$$R_{conv} = \frac{1}{hA} \tag{1.28}$$

Another important factor in convective heat transfer is friction coefficient (C_f), the characteristic of the fluid flow, which is:

$$C_f = \frac{\tau_w}{\rho_\infty u_\infty^2/2} \tag{1.29}$$

1.8.2.3 Radiation heat transfer

Radiation heat transfer is the energy that is emitted by matter in the form of photons or electromagnetic waves. Radiation can be important even in situations in which there is an intervening medium. An example is the heat transfer that take place between a living entity with its surrounding.

All bodies radiate energy in the form of photons. A photon is the smallest discrete amount of electromagnetic radiation (i.e., one quantum of electromagnetic energy is called a photon). Photons are massless and move in a random direction, with random phase and frequency. The origin of radiation is electromagnetic and is based on the Ampere law, the Faraday law and the Lorentz force. Maxwell analytically showed the existence of electromagnetic wave. Electromagnetic waves transport energy at the speed of light in empty space and are characterized by their frequency (v) and wavelength (λ) as follow:

$$\lambda = \frac{C}{v} \tag{1.30}$$

where C is the speed of light in the medium.

The electromagnetic waves appear in nature for wavelength over an unlimited range. Radiation with wavelength between 0.1 and 100 µm is in form of thermal radiation and is called radiation heat transfer. Thermal radiation includes the entire visible and infrared as well as a portion of ultraviolet radiation.

All bodies at a temperature above absolute zero emit radiation in all directions over a wide range of wavelengths. The amount of emitted energy from a surface at a given wavelength depends on the material, condition, and temperature of the body.

A surface is said to be diffuse if its surface properties are independent of direction and gray if its properties are independent of wavelength.

A black body is an ideal thermal radiator. It absorbs all incident radiation (absorptivity, $\alpha = 1$), regardless of wavelength and direction. It also emits maximum radiation energy in all directions (diffuse emitter). The energy radiated per unit area is given by Joseph Stefan as [51]:

$$\dot{Q}''_b = \sigma T^4 \tag{1.31}$$

Eq. (1.31) is called Stefan-Boltzmann law, σ is the Stefan-Boltzmann constant and is equal to 5.67×10^{-8} W/m^2 K^4, E_b is blackbody emissive power, and T is the absolute temperature. Real bodies radiate less effectively than black bodies. The rate of real body radiation energy per unit area is defined by [51]:

$$\dot{Q}'' = \varepsilon \dot{Q}''_b = \varepsilon \sigma T^4 \tag{1.32}$$

where ε is a property and is called the emittance. Values of emittance vary greatly for different materials. The emissivity of the human body is 0.97 for incident infrared radiation. They are near unity for rough surfaces such as ceramics or oxidized metals, and roughly 0.02 for polished metals or silvered reflectors.

Radiation energy can be absorbed, reflected, or transmitted when reaches a surface in human body. The sum of the absorbed, reflected, and transmitted fraction of radiation energy is equal to unity:

$$\alpha + \rho + \tau = 1 \tag{1.33}$$

where α is absorptivity (fraction of incident radiation that is absorbed), ρ is reflectivity (fraction of incident radiation that is reflected), and τ is transmissivity (fraction of incident radiation that is transmitted). Reflective energy may be either diffuse or specular (mirror-like). Diffuse reflections are independent of the incident radiation angle. For specular reflections, the reflection angle equals the angle of incidence.

1.8.3 Thermodynamic

Thermodynamics are consisted of two words: thermo (heat) and dynamics (power). It is a branch of science, deals with conversion of heat to work. It is established in the 19th century [52]. Historically, it dealt only with work generated by hot body (heat engine) and efforts to make it a more efficient heat engine. Today, thermodynamics deal mostly with energy and its relationship between properties of substances.

Thermodynamics generally starts with several basic concepts and lead to different thermodynamics laws. A thermodynamic system is a quantity of matter, which is defined by its boundary. Everything outside the boundary is called the surroundings or environment. The environment often contains one or more idealized heat reservoirs—heat sources with infinite heat capacity enabling them to give up or absorb heat without changing their temperature. The boundary can be real or imaginary, fixed or movable. There are two types of systems: closed and open system. Closed system (control mass) is a system with fixed quantity of matters. Thus, no mass crosses the boundary of the system. In open system (control volume) quantity of mass is not constant and mass can cross the boundary. Open system exchanges both matter and energy with its surroundings while the closed system only exchanges energy with its surroundings. The isolated system exchanges neither energy nor matter with its surroundings. An example of a true isolated system is the universe with of energy stored in it, is an isolated system.

Each system is characterized by its properties. Thermodynamic properties are a macroscopic characteristic of a living entity to which a numerical value is assigned at a given time without knowledge of its *history. Properties* are either intensive (exist at a point in space, like temperature, pressure, and density) or extensive [depends on the size (or extent) of the system, like mass and volume]. There are a number of different intensive properties that are used to characterize material behavior. The three most important independent properties that usually describe a system are temperature, pressure, and specific volume [52]. Temperature is the measure of the relative warmth or coolness of a body. In another word, it is the intensity of heat in an object and is expressed mainly by a comparative scale and shown by a thermometer. Pressure is the amount of force that is exerted on a surface per unit area and specific volume is the number of cubic meters occupied by 1 kg of a particular substance.

The four laws of thermodynamics that describe the temperature (zeroth law of thermodynamics), energy (first law of thermodynamics), entropy (second law of thermodynamics) and entropy of substances at the absolute zero temperature (third law of thermodynamics) are discussed in detail in the following sections.

1.8.3.1 Zeroth law of thermodynamics

This law serves as a basis for the validity of temperature measurement. The zeroth law of thermodynamics indicates that if two bodies are in thermal equilibrium with a third body, they are also in thermal equilibrium with each other. Replacing the third body with a thermometer helps measuring temperature of a system. Temperature is measured by means of a thermometer or other instrument having a scale calibrated in units called degrees. The size of a degree depends on the particular temperature scale is being used.

1.8.3.2 The first law of thermodynamics

Based on experimental observation, the energy can neither be created nor destroyed; it can only change forms. The first law of thermodynamics (or the conservation of energy principle) states that during an interaction between a system and its surroundings, the amount of energy gained by the system must be exactly equal to the amount of energy lost by the surroundings.

For a closed system (control mass), the first law of thermodynamics is shown as:

$$\left(\begin{array}{l}\text{Net amount of energy transfer as heat}\\ \text{and work to/or from the system}\end{array}\right) = \left(\begin{array}{l}\text{Net change in amount of enery}\\ \text{(increase/or decrease)with in the system}\end{array}\right)$$

or

$$Q - W = \Delta E \qquad (1.34)$$

In Eq. (1.34) Q and W is the net energy transfer by heat and work across the system boundary, respectively. The unit for Q and W is Joule (J). ΔE is the net change of total energy within the system and its unit is Joule (J). Total energy, E, of a system consists of internal energy (U), kinetic energy (KE), and potential energy (PE). The change in total energy of a system is expressed by:

$$\Delta E = \Delta U + \Delta KE + \Delta PE \tag{1.35}$$

Where:

$$\Delta U = \left(U_2 - U_1\right)\Delta KE = \frac{1}{2}m\left(V_2^2 - V_1^2\right)\Delta PE = mg\left(z_2 - z_1\right) \tag{1.36}$$

Internal energy is represented by the symbol U, and $U_2 - U_1$ is the change in internal energy of a process. V is the magnitude of system velocity, $\frac{1}{2}mV^2$ is the kinetic energy, KE, of the body and ΔKE is the change in kinetic energy of the system. Z is the magnitude of the system elevation relative to surface earth, mgz is the gravitational potential energy, PE, and the change in gravitational potential energy is ΔPE.

By inserting Eqs. (1.35), (1.36) into Eq. (1.34), the first law of thermodynamics becomes:

$$Q - W = \Delta U + \frac{1}{2}m\left(V_2^2 - V_1^2\right) + mg\left(z_2 - z_1\right) \tag{1.37}$$

The instantaneous time rate form of the first law of thermodynamics is:

$$\frac{dE}{dt} = \dot{Q} - \dot{W} \tag{1.38}$$

where $\dot{Q}$ and $\dot{W}$ are the rate of heat and work transfer across the boundary, respectively.

The first law of thermodynamics for open system (control volume) is expressed as:

$$\begin{pmatrix} \text{Time rate of change of the energy} \\ \text{within the control volume} \end{pmatrix} = \begin{pmatrix} \text{Net rate of energy transfer} \\ \text{as heat and work to/from} \\ \text{control volume at time } t \end{pmatrix} + \begin{pmatrix} \text{Net rate of energy transfer} \\ \text{by mass entering the} \\ \text{control volume} \end{pmatrix}$$

or

$$\frac{dE_{cv}}{dt} = \dot{Q}_{cv} - \dot{W}_{cv} + \sum_i \dot{m}_i \left(h_i + \frac{V_i^2}{2} + gz_i \right) - \sum_e \dot{m}_e \left(h_e + \frac{V_e^2}{2} + gz_e \right) \tag{1.39}$$

where subscripts i and e denote the inlet to and exit from the system, respectively, h is the enthalpy, and $\dot{m}$ is the mass flow rate.

1.8.3.3 The second law of thermodynamics

Second law of thermodynamics talks about the usefulness of energy as well as energy transfer direction. In another word second law of thermodynamics put limit on the first law of thermodynamics. Second law states that the total system work is always less than the heat supplied into the system. For closed system, the second law of thermodynamics is expressed as:

$$S_2 - S_1 = \int_1^2 \left(\frac{\delta Q}{T} \right)_b + S_{gen} \tag{1.40}$$

In Eq. (1.40) "b" is the system boundary, T is absolute temperature, $\dot{Q}$ is the rate of energy transfer by heat, and S_{gen} is the amount of entropy generated by system irreversibility. The irreversibility associated with a process (I) may be expressed by:

$$I = T_0 S_{gen} \tag{1.41}$$

The destruction of exergy due to irreversibility within the system is I and T_0 is temperature of the surroundings. The second law of thermodynamics for steady-flow system may be expressed as:

$$\frac{dS_{cv}}{dt} = \sum_j \frac{\dot{Q}_j}{T_j} + \sum_i \dot{m}_i s_i - \sum_e \dot{m}_e s_e + S_{gen} \tag{1.42}$$

Entropy in the microscopic or statistical view is a logarithm measure of the number of states (X_i) with significant probability of being occupied as is given as:

$$S = \sigma_B \ln X_i \tag{1.43}$$

σ_B is the Boltzmann constant (= 1.38×10^{-23} m^2 kg s^{-2} K^{-1}) [53].

1.8.3.4 The third law of thermodynamics

The third law of thermodynamics is formulated by Walter Nernst, also known as the Nernst heat theorem, and is based on the studies of chemical reactions at low temperatures and specific heat measurements at temperatures approaching absolute zero. The third law of thermodynamics states that the entropy of substances is zero at the absolute zero of temperature. An example is pure crystalline substance that has zero entropy at the absolute zero of temperature, 0 K.

References

[1] R. Siegel, D. Naishadham, A. Jemal, Cancer statistics, CA Cancer J. Clin. 62 (1) (2012) 10–29.
[2] L.S. Ries, M.A. Smith, J.G. Gurney, Cancer Incidence and Survival Among Children and Adolescents: United States SEER Program 1975–1995, National Cancer Institute, Bethesda, (1999) Publ No 99-4649.
[3] N. Howlader, A.M. Noone, M. Krapcho, et al. SEER Cancer Statistics Review, 1975–2008, National Cancer Institute, Bethesda, (2011).
[4] R.L. Siegel, K.D. Miller, A. Jemal, Cancer statistics, 2019, CA Cancer J. Clin. 69 (1) (2019) 7–34.
[5] L. Ries, M.P. Eisner, C.L. Kosary, et al. SEER Cancer Statistics Review, 1975–2002, National Cancer Institute, Bethesda, (2005) based on November 2004 SEER data submission http://seer.cancer.gov/csr/1975_2002/.

[6] P.J. Lachmann, Tumour immunology: a review, J. R. Soc. Med. 77 (12) (1984) 1023–1029.

[7] P. Friedl, S. Alexander, Cancer invasion and the microenvironment: plasticity and reciprocity, Cell 147 (5) (2011) 992–1009.

[8] F. Li, B. Tiede, J. Massagué, Y. Kang, Beyond tumorigenesis: cancer stem cells in metastasis, Cell Res. 17 (1) (2007) 3–14.

[9] M. Shackleton, F. Vaillant, K.J. Simpson, J. Stingl, G.K. Smyth, M.L. Asselin-Labat, L. Wu, G.J. Lindeman, J.E. Visvader, Generation of a functional mammary gland from a single stem cell, Nature 439 (7072) (2006) 84–88.

[10] C.F. Kim, E.L. Jackson, A.E. Woolfenden, S. Lawrence, I. Babar, S. Vogel, D. Crowley, R.T. Bronson, T. Jacks, Identification of bronchioalveolar stem cells in normal lung and lung cancer, Cell 121 (6) (2005) 823–835.

[11] T. Miyamoto, I.L. Weissman, K. Akashi, AML1/ETO-expressing nonleukemic stem cells in acute myelogenous leukemia with 8;21 chromosomal translocation, Proc. Natl. Acad. Sci. U. S. A. 97 (13) (2000) 7521–7526.

[12] A.L. Welm, S. Kim, B.E. Welm, J.M. Bishop, MET and MYC cooperate in mammary tumorigenesis, Proc. Natl. Acad. Sci. U. S. A. 102 (12) (2005) 4324–4329.

[13] J. Houghton, C. Stoicov, S. Nomura, A.B. Rogers, J. Carlson, H. Li, X. Cai, J.G. Fox, J.R. Goldenring, T.C. Wang, Gastric cancer originating from bone marrow-derived cells, Science 306 (5701) (2004) 1568–1571.

[14] S. Jaiswal, D. Traver, T. Miyamoto, K. Akashi, E. Lagasse, I.L. Weissman, Expression of BCR/ABL and BCL-2 in myeloid progenitors leads to myeloid leukemias, Proc. Natl. Acad. Sci. U. S. A. 100 (17) (2003) 10002–10007.

[15] L. Li, W.B. Neaves, Normal stem cells and cancer stem cells: the niche matters, Cancer Res. 66 (9) (2006) 4553–4557.

[16] L. Harrington, Does the reservoir for self-renewal stem from the ends?, Oncogene 23 (43) (2004) 7283–7289.

[17] J. Houghton, A. Morozov, I. Smirnova, T.C. Wang, Stem cells and cancer, Semin. Cancer Biol. 17 (3) (2007) 191–203.

[18] P.M. Biggs, Oncogenesis and herpes virus II, Part 2, in: G. de The, M.A. Epstein, H. zur Hausen (Eds.), Epidemiology, Host Response and Control II, IARC Publications, Lyon, 1972, pp. 317.

[19] R.T. Prehn, The immune reaction as a stimulator of tumor growth, Science 176 (4031) (1972) 170–171.

[20] S.I. Schlager, S.H. Ohanian, T. Borsos, Correlation between the ability of tumor cells to resist humoral immune attack and their ability to synthesize lipid, J. Immunol. 120 (2) (1978) 463–471.

[21] A. Bleyer, A. Viny, R. Barr (Eds.), Cancer Epidemiology in Older Adolescents and Young Adults 15 to 29 Years of Age, Including SEER Incidence and Survival: 1975–2000, National Cancer Institute, Bethesda, 2006 NIH Pub. No. 06-5767.

[22] M.F. Greaves, Analysis of the clinical and biological signify cance of lymphoid phenotypes in acute leukemia, Cancer Res. 41 (11 Pt 2) (1981) 4752–4766.

[23] J.R. Wilczynski, M. Duechler, How do tumors actively escape from host immunosurveillance?, Arch. Immunol. Ther. Exp. (Warsz.) 58 (6) (2010) 435–448.

[24] M. Lewis, G. Lewis, Strengthen the body and its immune cells, in: M. Lewis, G. Lewis (Eds.), Cancer—A Threat to Your Life OR A Chance to Control Your Future, Australia Cancer & Natural Therapy Foundation of Australia; Lewis Publications, New Zealand, 2006.

[25] G. Szekeres, Z. Battyani, Immuno-diagnosis of malignant melanoma, Magy Onkol. 47 (1) (2003) 45–50.

[26] W.B. Coley, The treatment of inoperable sarcoma by bacterial toxins (the Mixed Toxins of the *Streptococcus erysipelas* and the *Bacillus prodigiosus*), Proc. R. Soc. Med. 3 (Surg Sect) (1910) 1–48.

[27] W.J. Lesterhuis, J.B. Haanen, C.J. Punt, Cancer immunotherapy—revisited, Nat. Rev. Drug Discov. 10 (8) (2011) 591–600.

[28] P.S. Moore, Y. Chang, Why do viruses cause cancer? Highlights of the first century of human tumour virology, Nat. Rev. Cancer 10 (12) (2010) 878–889.

[29] H.L. Kaufman, D.W. Kim, G. DeRaffele, J. Mitcham, R.S. Coffi n, S. Kim-Schulze, Local and distant immunity induced by intralesional vaccination with an oncolytic herpes virus encoding GM-CSF in patients with stage IIIc and IV melanoma, Ann. Surg. Oncol. 17 (3) (2010) 718–730.

[30] K.S. Peggs, S.A. Quezada, C.A. Chambers, A.J. Korman, J.P. Allison, Blockade of CTLA-4 on both effector and regulatory T cell compartments contributes to the antitumor activity of anti-CTLA-4 antibodies, J. Exp. Med. 206 (8) (2009) 1717–1725.

[31] A. van Elsas, A.A. Hurwitz, J.P. Allison, Combination immunotherapy of B16 melanoma using anticytotoxic T lymphocyte-associated antigen 4 (CTLA4) and granulocyte/macrophage colony-stimulating factor (GM-CSF)-producing vaccines induces rejection of subcutaneous and metastatic tumors accompanied by autoimmune depigmentation, J. Exp. Med. 190 (3) (1999) 355–366.

[32] W. Zou, L. Chen, Inhibitory B7-family molecules in the tumor microenvironment, Nat. Rev. Immunol. 8 (6) (2008) 467–477.

[33] R.H. Thompson, S.M. Kuntz, B.C. Leibovich, H. Dong, C.M. Lohse, W.S. Webster, et al. Tumor B7-H1 is associated with poor prognosis in renal cell carcinoma patients with long-term follow-up, Cancer Res. 66 (7) (2006) 3381–3385.

[34] J. Dannull, Z. Su, D. Rizzieri, B.K. Yang, D. Coleman, D. Yancey, et al. Enhancement of vaccine-mediated antitumor immunity in cancer patients after depletion of regulatory T cells, J. Clin. Invest. 115 (12) (2005) 3623–3633.

[35] M.E. Dudley, J.R. Wunderlich, P.F. Robbins, J.C. Yang, P. Hwu, D.J. Schwartzentruber, et al. Cancer regression and autoimmunity in patients after clonal repopulation with antitumor lymphocytes, Science 298 (5594) (2002) 850–854.

[36] M. Obeid, A. Tesniere, F. Ghiringhelli, G.M. Fimia, L. Apetoh, J.L. Perfettini, et al. Calreticulin exposure dictates the immunogenicity of cancer cell death, Nat. Med. 13 (1) (2007) 54–61.

[37] W.J. Lesterhuis, I.J. de Vries, E.A. Aarntzen, A. De Boer, N.M. Scharenborg, M. Van de Rakt, et al. A pilot study on the immunogenicity of dendritic cell vaccination during adjuvant oxaliplatin/capecitabine chemotherapy in colon cancer patients, Br. J. Cancer 103 (9) (2010) 1415–1421.

[38] M.H. Den Brok, R.P. Sutmuller, R. Van der Voort, E.J. Bennink, C.G. Figdor, T.J. Ruers, et al. In situ tumor ablation creates an antigen source for the generation of antitumor immunity, Cancer Res. 64 (11) (2004) 4024–4029.

[39] P.M. Brickell, D.S. Latchman, D. Murphy, K. Willison, P.W. Rigby, Activation of a Qa/Tla class I major histocompatibility antigen gene is a general feature of oncogenesis in the mouse, Nature 306 (5945) (1983) 756–760.

[40] H.D. Shukla, Comprehensive analysis of cancer-proteogenome to identify biomarkers for the early diagnosis and prognosis of cancer, Proteomes 5 (4) (2017) 28.

[41] Shruthi, B.S., Vinodhkumar, P., & Selvamani. (2016). Proteomics: A new perspective for cancer. Adv. Biomed. Res., 5, 67.

[42] G.A. Calin, C.M. Croce, MicroRNA signatures in human cancers, Nat. Rev. Cancer 6 (11) (2006) 857.

[43] T. Walsh, M.K. Lee, S. Casadei, A.M. Thornton, S.M. Stray, C. Pennil, et al. Detection of inherited mutations for breast and ovarian cancer using genomic capture and massively parallel sequencing, Proc. Natl. Acad. Sci. 107 (28) (2010) 12629–12633.

[44] J. Wallyn, N. Anton, S. Akram, T.F. Vandamme, Biomedical imaging: principles, technologies, clinical aspects, contrast agents, limitations and future trends in nanomedicines, Pharm. Res. 36 (6) (2019) 78.

[45] Higgins, L. J., & Pomper, M. G. (2011, February). The evolution of imaging in cancer: current state and future challenges. In Seminars in oncology (Vol. 38, No. 1, pp. 3-15). WB Saunders, Boston, USA.

[46] Y. Bayazitoglu, S. Kheradmand, T.K. Tullius, An overview of nanoparticle assisted laser therapy, Int. J. Heat Mass Transfer 67 (2013) 469–486.

[47] R.W. Fox, A.T. McDonald, P.J. Pritchard, Introduction to Fluid Mechanics, fifth ed. John Wiley & Sons, Inc., 2010.

[48] L. Waite, J.M. Fine, Applied Biofluid Mechanics (2007).

[49] G. Falkovich, Fluid Mechanics: A Short Course for Physicists, Cambridge University Press, Cambridge, United Kingdom, (2011).

[50] J.R. Womersley, Method for the calculation of velocity, rate of flow and viscous drag in arteries when the pressure gradient is known, Jo. Physiol. 127 (3) (1955) 553–563.

[51] J.P. Holman, Heat Transfer, McGraw-Hill, New York, NY, (2010).

[52] Y.A. Cengel, M.A. Boles, Thermodynamics: an engineering approach, Sea 1000 (2002) 8862.

[53] K. Stowe, An introduction to Thermodynamics and Statistical Mechanics, Cambridge University Press, Cambridge, United Kingdom, (2007).

Diagnostic imaging in cancer

Chapter outline

2.1 X-ray-based systems including CT scan

In general, the radiography technic includes all types of X-ray scanners. While X-ray computed tomography (X-ray CT) and microcomputed tomography (micro-CT scan) are dedicated to clinical and preclinical studies, respectively. Micro-CT scan is divided into three category including mini-CT scan (with in vivo scale of 50–200 µm), micro-CT scan (with in vivo scale of 1–50 µm), and nano-CT scan (with in vivo scale of 0.1–1 µm), as shown in Fig. 2.1A. Generally, preclinical studies are applied micro-CT to examine small scale animals such as mice [1].

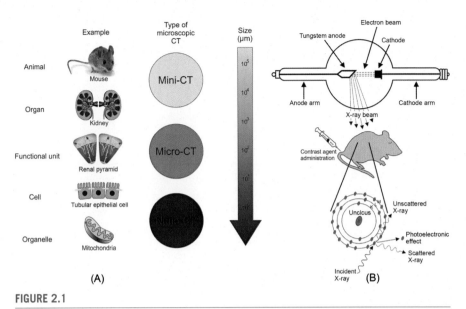

FIGURE 2.1

(A) Types of micro-CT with their in vivo scale. (B) Schematic of the X-ray scanner principle.

2.1.1 Principle

X-ray imaging is based on the ability of desired tissue to attenuate the emitted X-ray resulting the formation of an image with different opacities. In general, X-ray image is formed according to the following protocol, which are represented in Fig. 2.1B.

1. X-ray is generated by an electron beam accelerated by the electron filed and guided to the heavy metal anode.
2. X-ray passes through the desired specimen.
3. X-ray diffuses to the specimen based on the type of molecules.
4. The X-ray interaction with molecules result reflection, absorption, or scattering of the emitted photons. X-ray absorption follows Beer-Lambert law depending on the electron density, the energy of the photons, the thickness of the objects, and the absorption coefficient of the element in in vivo media [2].
5. Total attenuation is obtained by analysis the emergent X-ray photons. Tissue opacification is visualized by reconstruction of sectional image to form 2D and 3D images.

The image contrast is measured by grayscale which depends on the material density. Dense materials absorb a large amount of X-ray photons resulting the formation of white images (white to light gray) while fluids absorb less photons leading to the formation of dark images (dark gray to black).

Opacification is quantified using Hounsfield unit (HU). Based on this scale, air and water possess the values of -1000 and 0 HU, respectively. HU value for soft tissues is mainly between -100 and $+100$. Mineralized materials are typically enjoying a value between +400 and +1000 HU. HU value is determined according to the following equation [3]:

$$HU = 1000 \times \left(\frac{\mu - \mu_{water}}{\mu_{water}} \right)$$
(2.1)

where μ and μ_{water} are the X-ray attenuation coefficient of the desired target and water, respectively.

One of the main limitations of this technic is related to the low difference between HU value of soft tissues making the imaging process difficult. To tackle the problem, radiopaque contrast agents (CAs) are introduced which improve the sensitivity of X-ray scanner.

2.1.2 Kinds of radiopaque contrast agent

To increase the accuracy of image obtained by X-ray, radiopaque CAs are administrated during imaging process. By increasing the amount of radiopaque CAs, the contrast of image is increased in a desired region.

X-ray attenuation may be achieved by using heavy elements as a radiopaque CAs as summarized in Fig. 2.2. Many radioactive elements such as thorium [4] is withdrawn from clinical application due to its cytotoxicity. Also, for patients suffering from severe diabetes, renal failure and iodine sensitivity, and radiopaque materials are contraindicated [5]. Nowadays, oral barium sulfate and iodinated molecules are the most used radiopaque materials. Salt such as lithium iodine and sodium iodine are not used as a radiopaque anymore due to the charge separation in in vivo media. Also, several studies have been focused on the synthesis of various formulation of iodine with different osmolarities. Currently, low-molecular weight iodinated molecules containing triiodobenzene groups are the best radiopaque CAs since has the highest benefit-to-risk profile [6,7]. Commercial iodinated molecules for clinical application as a radiopaque CAs are include Iopromide (Ultravist), Iopamidol (Isovue), Iohexol (Omnipaque), and Iodixanol (Visipaque). Although ionic molecules improve cellular uptake via the interaction between positively charged molecules and negatively charged cellular membrane, they suffer from toxicity. One of the main limitations of using radiopaque CAs are related to the short half-life in blood circulation of

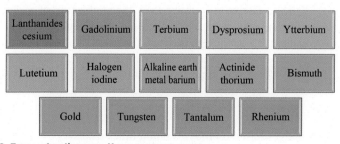

FIGURE 2.2 Types of radiopaque X-ray contrast agent.

hydrophilic molecules due to the fast-renal excretion. In order to improve the sensitivity of X-ray scanners, high dose molecules need to be administrated which may cause acute renal toxicity, cardiovascular problems, and allergy. To deal with the problem, a new class of lipid-based radiopaque CT has been introduced to replace hydrophilic iodinated CAs such as Lipiodol [7] and Fenestra [8]. Furthermore, nanotechnology brings new types of radiopaque CT to the market [9].

Recent progress in technology develops new applications for the use of CT, such as parasite imaging [10], coronary inflammation [11] and determining the initial breast tumor size [12].

2.1.3 Advantages and disadvantages

The main consideration of using X-ray imaging technic is related to an assessment of benefit-to-risk profile.
The key advantages of this technic include:

1. Easy and quick processing
2. Having high resolution
3. Be able to diagnosis various kinds of diseases including broken bone, infection, and cancer

The main limitation of this technic ise as follow:

1. Causing genetic damage due to the delivery of high dose
2. Causing nephropathy and cardiovascular problems
3. Inducing cancer or increasing cancer life time
4. Costly

It should be noted that all these side effects depend on the amount of photon energy, number of scan and scan speed [13]. The damages are minimized by compromising the quality of the image and patient health.

2.2 Magnetic resonance systems

For many decades, there has been an opinion that magnetic field passes through the body with no harmful effects [14]. Since then, the application of magnetic field in biomedicine has been introduced. Magnetic resonance imaging (MRI) is a nonionizing imaging technic providing high resolution 3D-images of soft tissues. Currently, MRI is one of the most well-known tools for human visualization [15]. It mostly has been used for tumors detection, brain, and cardiovascular observation.

2.2.1 Principle

MRI provides images based on the relaxation times (transvers T_2 and longitudinal T_1) of water in a magnetic field. In body, each compartment has been surrounded

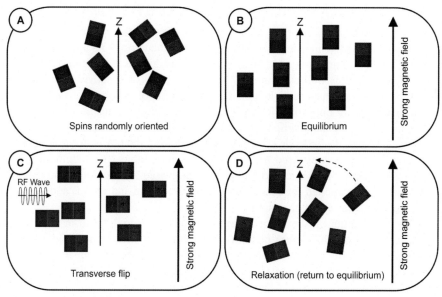

FIGURE 2.3 MRI principle.

by water molecules with specific relaxation time [16]. Differences in relaxation time are related to the density, composition of each organ and region surrounded organ. Fig. 2.3 represents the general principle of four steps of MRI. The first step relies on the response of protons from water to the magnetic field ($B_0 > 2$ T). Although magnetic moment of each hydrogen molecule is small, the presence of large amount of water molecules in the biological media leads to the formation of considerable net magnetic moment from all of the hydrogen molecules. In the second step, radiofrequency (RF) pulses (5–100 MHz) are applied to the aligned proton nuclei. In the third step, the aligned protons which are excited by RF, flip to the transverse plan. Finally, by ending RF pulses, net magnetization of hydrogen molecules become relax and hydrogen molecules are aligning along magnetic field. The relaxation time is treated by Fourier transform and processes to create 3D images [16].

Relaxation is based on two mechanisms (1) longitudinal relaxation related to the transferring of absorbed energy to the surrounded media and (2) transversal relaxation related to the loss of phase coherence during spin processing. In detail, T_1 and T_2 describe the time to relax 63% of longitudinal magnetization and 37% of transversal magnetization, respectively [17]. T_1 recovery and T_2 decay provide brightening and darkening contrast, respectively. Another considerable phenomenon is the dephasing mechanism occurring due to the magnetic field inhomogeneity in the tissue named T_2^*. Generally, T_2^*-weighted imaging is detected during T_2-weighted imaging. To improve image quality, magnetic materials have been administrated to enhance image contrast.

2.2.2   Kinds of magnetic probes to enhance MRI contrast

In general, there are two types of MRI CA, one of the groups is related to the enhancing T_1-weighted MRI and the other is related to the enhancing of T_2- and T_2^*-weighted MRI [18]. Two main types of MRI CA are based gadolinium and iron oxide for T_1-weighted imaged as well as T_2 and T_2^*-weighted image, respectively. Both categories are well-coated nanoparticles [2]. The application of nanoparticles as a CA has been increased not only for MRI, for the other imaging systems as well. The main advantages of using nanoparticles as a contest agent are increasing blood circulation time, targeting to desired side of action, and increasing the CA concentration in the targeted tissue. The most used MRI CAs is shown in Table 2.1. Gadolinium-based CAs mainly are divided into two groups: (1) intravascular, interstitial and extracellular fluid agents applied for examining lymphatic system and vessels and (2) blood pool MRI CA used for angiography.

Since gadolinium in ionic form has undesirable cell toxicity in in vivo media, considerable efforts have been done to connect this ion with ligands. Gd^{3+} chelates possess high thermodynamic stability and high kinetic, improving the images of interstitial spaces. Furthermore, Gd^{3+} chelates cannot pass through blood brain barrier and mostly are administrated to improve the CA of blood vessels and tumors. Due to their hydrophilic nature, Gd^{3+} chelates filter by the kidney and undergo renal clearance. It should be noted that transition elements (such as Mn^{2+}, Fe^{3+}, and Cu^{2+}) are used as a CA for MRI due to having a large number of unpaired electrons, suitable for longitudinal MRI imaging. So far, gadolinium-based CAs are the most used CA for positive imaging in MRI. Based on the application, there are large variety of CAs from ionic to nonionic and linear to nonlinear with different ratios of Gd^{3+} to ligand.

Table 2.1 Types of MRI contrast agent.

MRI contrast enhancer	
Gadolinium	**Iron oxide**
Gadopentetate dimeglumine (Magnevist)	Ferumoxsil (Lumirem or GastroMark)—silicon coating
Gadoterate meglumine, (Dotarem)	Ferristene (Abdoscan)—sulfonated styrene-divinylbenzene copolymer citrate coating
Gadoteridol (ProHance)	Ferumoxide (Endorem)—dextran coated-Fe3O4
Gadodiamide (Omniscan)	Ferucarbotran (Resovist)—carboxyldextran coating
Gadobutrol (Gadovist)	Ferumoxtran-10 (Sinerem or Combinex)—dextran coating
Gadobenate dimeglumine (MultiHance)	Feruglose (Clariscan)—PEG starch coating
Gadoversetamide (OptiMark)	
Gadoxetic acid (Primovist or Eovist)	
Gadofosveset (Vasovist)	

The efficiency of MRI CAs is related to their impact on the relaxation time. The efficiency of CAs defines as r_1 and r_2 (m Mol^{-1} s^{-1}) related to the longitudinal and transversal relaxation, respectively:

$$r_1 = \frac{1}{T_1} \quad \text{and} \quad r_2 = \frac{1}{T_2} \tag{2.2}$$

The r_2/r_1 ratio is evaluated to determine the efficiency of CA; as the ratio of r_2/r_1 increase, the contrast of T_2-weighted imaging is increased. Experimentally, r_1 and r_2 are calculated using the plot of $1/T_1$ and $1/T_2$ versus iron concentration by applying the following equation:

$$\frac{1}{T_i} = \frac{1}{T_i^0} + r_i C \quad i - 1,2 \tag{2.3}$$

Where T_1^0 and T_2^0 are the longitudinal and transversal relaxation time in pure water, respectively and C is the concentration of iron oxide [16,19].

2.2.3   Advantages and disadvantages

The advantages of MRI are as follow:

- No need for iodinated contrast
- Greater ability to image the brain and spinal cord than other modalities such as CT
- Nonradiation involved
- Excellent contrast for soft tissues
- Allow lesion detection
- Staging of cancer is possible

The possible disadvantages of MRI are listed as follow:

- Expensive
- Not portable
- Difficult for real time imaging
- Limited availability
- Lengthy acquisition time
- Limited information on lung
- No electron density mapping for the tissues
- Contraindicated in some ferromagnetic implant such as pacemaker, fusion, and screws

2.3   Ultrasound

Ultrasound is longitudinal acoustic energy having a frequency above 20 kHz (i.e., above human hearing range). Ultrasound irradiation makes particles displacement along with the wave propagation. Currently, ultrasound imaging is the most clinically used technic after CT scan [20].

The basic steps of ultrasound imaging are:

1. Ultrasound wave is generated by ceramic probe containing a crystal. The main mechanism of creating ultrasound wave relies on the piezoelectric effect which converts electrical energy to the mechanical energy makes the crystal to vibrate.
2. The wave is passed through the specimen and partially reflected at each tissue interface (between fluid and soft tissue as well as soft tissue and bone). All of the sound waves are undergoing the following phenomenon during propagation inside the specimen: refraction, diffraction, scattering, and attenuation.
3. The reflected sounds are picked up by the piezoelectric crystal. In this step, mechanical waves are transferred to electrical energy.
4. The distance between the boundaries in the tissue and the scan probe is calculated using speed of sound and then converted to the signals in the machine displays.

The large variety of applications has been considered for ultrasound imaging, including vascular diagnosis, tumor examination, following organ after transplantation.

Beside diagnostic interests, high intensity focused ultrasound (HIFU) has been absorbed a great deal of attention due to the valuable therapeutic potential for destroying kidney stone [21], benign and malignant tumor [22]. In this approach, the ultrasound waves are focused on the pathological tissue with high energy and then the acoustic energy is absorbed by targeted tissue. So the temperature of the tissue is increased leading to destroyed the pathologic tissue.

To improve ultrasound imaging features, ultrasound contrast imaging which composed of gas-filled microbubbles surrounded by a lipid-based shell has been introduced. Ultrasound contrast imaging examines pathological tissues regarding functional and molecular vascular properties.

One of the main applications of ultrasound CA is related to the liver examination due to the considerable improvement in compared to conventional ultrasound imaging which may lead to the detection of malignant liver in early stage and rapid washout of liver CA [23]. Furthermore, ultrasound CA is administrated for the patients having contra-indications to MRI or CT contrast-enhancer due to the sever renal dysfunction.

The efficiency of CA is examined based on its sensitivity and specificity. Microbubbles-based CA has high sensitivity even a single of which can be detected. Also, due to the microbubbles size (1–5 μm) they do not extravasate from the vessels which has both advantages and disadvantages. On the one hand, targeting to interstitial space is completely feasible. On the other hand, extravascular targeting is hard to achieve due to their size.

Three-dimensional ultrasound consists of two main categories. One is the 2-D imaging equipment coupled with a specific mechanical movement to build 3-D imaging. The other is online monitoring of volumetric echo which composes of specific transducer for 3-D volume scanning.

2.3.1 Microbubbles-based contrast agent

The density of blood vessels in tumorogenic side is proportional to the size and pathological condition of the tumors. To visualize vascular morphology and tissue surrounded by different vascular morphology, contrast-enhanced ultrasound imaging has been administrated.

As already mentioned, these kinds of CAs are generally gas-filled microbubbles ranging 1–5 μm in diameter. The acoustic response of microbubbles is around MHz range which is optimal for ultrasound imaging [23]. Microbubbles are able to reduce backscattering from the vasculature.

In order to increase blood circulation time and improve the stability of microbubbles, a lipid shells such as lipid [24], polymers [25], protein [26], or a mixture of them [27] are surrounded them. Beside the shells, the type of encapsulated microbubbles affects the stability and in vivo half-life of the CA. It has been shown that low solubility of the encapsulated microbubble (Ex., perfluorochemicals) enjoys high stability and circulation time [28]. It would be worth noting that, the amount of shell's oscillating after insonication is determined. In this context to types of shells have been described: soft shells mainly used for hormonic imaging and hard shells applied for destructive ultrasound imaging. One of the commonly used microbubbles is SonoVue which is a perfluoro agent gas surrounded by a phospholipid shell.

The ultrasound molecular imaging involves the use of targeted microbubbles to selective bind to the desired region in intravascular space. Target-specific microbubbles mostly are achieved by attaching antibodies or peptides to the surface of microbubbles either during the production process of microbubbles or after its synthesize.

The main aims for target-specific microbubbles are molecules expressed on the activated endothelium in response to either angiogenic stimuli or inflammation. The main targeted molecules in this field of interest are including vascular cell adhesion molecule (VCAM-1), intercellular adhesion molecule (ICAM-1), interleukins (IL-1), tumor necrosis factor-α (TNF-α), vascular endothelial growth factor (VEGF), fibroblast growth factor (FGF), and endoglin.

Due to the difference in compressibility and acoustic impedance between the surrounding media and the microbubbles, contrast-enhanced ultrasound imaging mainly acts as nonlinear scatters. Nonlinear imaging technic, which includes subharmonic imaging and intermittent power Doppler, improves the quality of the images by reducing bubble destruction.

2.3.2 Advantages and disadvantages

The advantages of using ultrasound imaging are as follow:

- Being portable
- Easy to use
- Appropriate for soft tissue
- Cost-effective
- Extensive database for some species

- No radiation
- Can be applied for real-time analysis
- No size limitation
- Not require general anesthesia

The disadvantages of using ultrasound imaging are:

- Unsuitable for imaging lungs and bowel
- Resolution is not as good as CT or MRI
- Image analysis is not easily automated
- No whole-body information
- Target-specific imaging is limited to intravascular compartment

2.4 Nonionizing electromagnetic imaging

Nonionizing electromagnetic radiation refers to the low photon energy portion of the electromagnetic spectrum, from 1 to 3×10^{15} Hz. In comparison to ionizing radiation, which the interaction with atom result in remove electron, interaction of nonionizing electron leads to an excitation and heat production.

The penetration depth in human body, sites of absorption, and the consequent side effects of the nonionizing electromagnetic radiation are very much depending to the particular wavelength.

2.4.1 Kinds of nonionizing electromagnetic imaging

Electromagnetic radiation is divided into two categories: ionizing and nonionizing radiation (Fig. 2.4). Nonionizing electromagnetic imaging tools is mainly divided into three most studies categories include (1) thermo-acoustic imaging, (2) electrical impedance tomography (EIT), and (3) near-infrared (NIR) optical tomography.

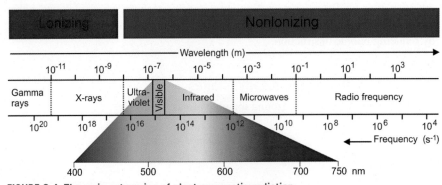

FIGURE 2.4 The main categories of electromagnetic radiation.

2.4.1.1 *Thermo-acoustic imaging*

Thermo-acoustic imaging systems use nonionizing radiation, such as RF, light absorption, and microwave to induce ultrasound waves in the targeted tissues. The basic steps of image formation in this technic are in five steps. First, the electromagnetic energy is emitted as uniformly as possible. Second, the deposited energy is absorbed by the tissue leading to increase the temperature (less than 0.001°C). Third, increase in temperature causes thermal expansion, however slightly. Fourth, this mechanical expansion leads to produce acoustic wave propagating in all directions and detect by transducer surrounding the target. Fifth, the transducers, which is mainly piezoelectric crystals, scan the target and collect the tomographic data. (The summary of image formation is shown in Fig. 2.5.)

Photo-acoustic is a type of thermoacoustic which the generating radiation is optical. Typically, the pulse duration in the photo-acoustic imaging is around 5–10 ns. Due to the increase in the amount of hemoglobin and water around a tumorogenic tissue, more absorption in electromagnetic energy and consequently more thermal expansion has occurred than surrounded healthy tissues [29]. Reports have been shown that photo-acoustic imaging with very short wavelengths has created a high contrast between tumorogenic and nontumorogenic tissues [30].

The features of emitted sources are different. For example, the penetration depth of RF and microwave are more than optical pulses, while microwave pulses are less uniform as compared with the others and are mainly used for preclinical studies [31].

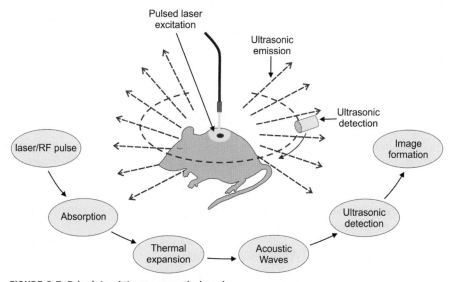

FIGURE 2.5 Principle of thermoacoustic imaging.

The guided equation for photo-thermoacoustic computed tomography (CT) is explained as follow:

$$\nabla^2 p(\vec{r},t) - \frac{1}{v_s^2}\frac{\partial^2}{\partial t^2} p(\vec{r},t) = -\frac{\beta}{C_p}\frac{\partial}{\partial x}H(\vec{r},t) \tag{2.4}$$

where $H(\vec{r},t)$ heating function, v_s is the speed of sound in the medium, C_p specific heat capacity, and β is thermal expansion coefficient. To assure that heat diffusion is negligible during pulse laser irradiation, the pulse laser frequency is chosen much shorter than the thermal relaxation time. $P(\vec{r},t)$ in the above equation is obtained as follow:

$$p(\vec{r},t) = \frac{\beta}{4\pi C_p}\int \frac{d\vec{r}'}{|\vec{r}-\vec{r}'|}\left.\frac{\partial H(\vec{r}',t')}{\partial t'}\right|_{t'=t-} |\vec{r}-\vec{r}'|/v_s \tag{2.5}$$

when laser pulse frequency is much shorter than laser relaxation time, the stress is confined and as a result, the above equation is rewritten as follow:

$$p(\vec{r},t) = \frac{1}{4\pi v_s^2}\frac{\partial}{\partial t}\left[\frac{1}{v_s t}\int d\vec{r}' p_0(\vec{r}')\delta\left(t - \frac{|\vec{r}-\vec{r}'|}{v_s}\right)\right] \tag{2.6}$$

where p_0 is the initial photoacoustic pressure.

2.4.1.2   *Electrical impedance tomography*

EIT is a kind of noninvasive medical imaging determines the electrical permittivity, conductivity and impedance of the interior part of the body from surface electrode. Conducting surface electrodes are attached to the skin and small alternating currents at a single or multiple frequency are applied to the electrodes, then the electrical potentials are measured and used to build a tomographic image of the target. EIT imaging technic is based on the measuring the tissue impedance consisting of membranes, cells, and fluids. Cells and membranes act as small imperfect capacitors with high resistivity, while fluids are the resistive component that contribute a frequency dependence for liquids in the outside of the cells. At high frequencies (in the range of MHz), the only resistive component is related to the conduction through extracellular and intracellular fluid, while at low frequencies (in the range of Hz to kHz), membranes impede the current and can be used to obtain valuable information regarding the size, morphology, and electrical properties of cells [32]. There are two types of EIG systems including absolute EIT and difference EIT. Difference EIT focuses on the study of the changes in conductivity during a period of time. Absolute EID is more difficult since channel noise and electrode impedance need to be calculated. One of the most clinical applications of EID for cancer diagnosis is related to the detection of malignant breast tumors from benign lesions [33].

2.4.1.3   *Near-infrared optical tomography*

NIR optical tomography is a noninvasive imaging technic with high contrast features for blood cells. This is mainly due to the fact that hemoglobin absorbs NIR light while water absorption is far from NIR wavelength. This technic provides valuable information regarding hemoglobin concentration, water percentage, lipid concentration, oxygen saturation, vascularity, and scattering related properties of the targeted tissue [34, 35]. NIR optical tomography can also be used for cancer detection in endoscopy [34].

Although NIR optical tomography enjoys high contrast imaging properties, it is restricted to tissue with the ability of transluminated externally. One of the excellent subjects for NIR optical tomography is breast tissues due to the direct contact with NIR detector [35]. Also, the high vasculature of internal tumors such as colon, rectum, and prostate provide high contrast for NIR optical tomography.

2.4.2   Advantages and disadvantages

The advantages of using nonionizing electromagnetic imaging system is described as follow:

- Cancer and other chronic disease treatment by combining chemotherapy with nonmagnetic pulse irradiation
- Diathermy by low-level radiation
- Not known genetic damage
- Cost-effectiveness
- Can scan in multiple plan
- Can scan in 1 mm to several centimeter

The disadvantages of using nonionizing electromagnetic imaging are summarized as fallow:

- Photo-aging of the skin
- Corneal injury
- Creating erythema
- Photosensitization
- Thermal retinal injury
- Corneal burn
- Changing the permeability of blood-brain barriers
- Changing hormone level

2.5   Radiopharmaceutical imaging

Radiopharmaceutical agents are a radioactive component containing radionuclides or radioisotopes. A radioisotope is an unstable atom achieving a more stable level by radiating energy in the form of a positron particle, beta particle, or gamma-ray. The releasing or radiating energy is achieved under three specific mechanisms include physical decay, disintegration, or transition. Releasing energy may either change the isotope of the atom but not the element, or completely change the element by transmutation [36]. Radiopharmaceuticals consist of a wide variety of different radioactive agents used for imaging, radiotherapy or both (theranostic purpose). These molecules inject intravenously and consider as relatively noninvasive methods.

Nuclear medicine mainly uses γ radiation for cancer imaging while α and β radiation mainly applied for radioimmunotherapy (RIT). Typical radionuclides used in cancer imaging include ^{131}I, ^{123}I, ^{67}Ga, ^{99m}Tc, ^{111}In, and ^{201}Tl [34]. The uptake and in vivo distribution of these agents are mainly depend on their pharmokinetic properties. It should be noted that simple chemistry of these molecules provides feasible opportunity for ligand conjugation on their surface leading to increase the chance of accumulation in desired region. Although the amounts of radioactive molecules are very small, around picogram range, which cannot affect the pharmaco distribution, the irradiated radioactive energy is sufficient for imaging and therapeutic purpose. Radiopharmaceutical agents mainly are conjugated to relatively high amount of biologically active ingredient being nonradioactive molecules.

After administration of radionucleotide, the patient is then scanned using single photon emission CT and planar gamma cameras. In this system valuable information regarding to the anatomical details, physiological and functional response can be obtained. Also, the radiopharmaceutical component in targeted tissue are used as a diagnostic biomarker. The diagnostic biomarker is designed with the special interest to the biomolecular targets in special tissue or physiological process. Finding biomolecules with specific interest to one particular biological molecule are of interest. However, radiolabeled antibodies or radiolabeled small molecules have been designed with very high affinity to few biomolecules.

Among all the other types of scanner (e.g., CT, MRI, and ultrasound), nuclear imaging scanners are time-consuming process. The total period of time is needed for nuclear imaging study depends on technical and biological varieties. After intravenously injection of radiopharmaceutical agent, the imaging process is started with necessarily delay to let the radiopharmaceutical to spread in the whole body. Then, for data acquisition, patient must spend sufficient time in front of camera to collect all the necessarily radioactive signals. The amount of time needed for collecting data is mainly related to the intrinsic properties and concentration of administrated radioisotope in the region of interest, the accuracy of camera system as well as the way of image reconstruction [36].

2.5.1 Application of radiopharmaceutical imaging

There are large verity of applications for radiopharmaceutical imaging including bone scanning, lymphoscintigraphy, immunoscintigraphy, RIT, peptide receptor radionuclide therapy (PRRT), scintimammography, angiogenesis imaging, and multidrug resistance imaging.

One of the main applications of nuclear medicine has been focused on the bone scanning due to the good sensitivity and low cost. Technetium-based radiopharmaceuticals and fluorodeoxyglucose-based positron emission tomography (FDG-PET) have been used to detect metastases [37]. Another application of nuclear medicine is lymphoscintigraphy. Lymph node drainage has been studied by human serum albumin labelled with ^{99m}Tc. The first lymph node draining a cancer is sentinal lymph node [38]. The absence of cancer cells in sentinal lymph node represent the absence of metastatic cancer cells in all the other nodes. Lymphoscintigraphy can prevent unnecessarily surgery if the sentinal lymph node is found to be free of cancer cells. One of the important parameters for well-characterized lymphatic drainage is the size of radiocolloids which is inversely proportional to the draining [34]. As already mentioned, immunoscintigraphy applies antibody to target specific antigens. One of the well-studied cancer with immunoscintigraphy is prostate. ^{111}In was labeled with capromab pendetide which is binding to the prostate membrane specific antigen.

Radiopharmacutics use not only for imaging but also for therapy. RIT and PRRT can specifically irradiate cancer cells. These systems involve low-dose irradiation from radionuclides targeted tumor. The therapeutic effects lead to the absorption of energy from radionuclide's emissions. Over expression of peptides involved in proliferation and angiogenesis is used for labeling radionecleotides for investigation of receptor-positive tumors. Radiolabeled bombasin, somatostatin and neuropeptide Y are used in scintigraphy to examine receptor-positive cancer cells. One of the most studied radiolabeled nucleotide in this field of interest is ^{111}In-DTPA-octeotride used for somatostatin targeting scintigraphy imaging. Also, this radionucleotide has been labeled for α and β radionuclides therapy for PRRT. Another examples are the use of ^{90}Y-DOTATOC and b-particle for radionuclides therapy [34]. Another application of radionucleotides is for scintimammography. There is a need to develop a gamma camera with the ability of reducing the scattering from external sources specially when the image is taken near the chest wall. Also, designing a camera with improving image resolution at the edge of detector will improve the quality of images taken from breast tissue near to the chest wall [34]. Based on the special characteristic features of breast tissue, breast specific gamma camera imaging is developed [39, 40]. This new type of camera possess small field of view which increases the image resolution, improves flexibility of movement and applies compression to the breast to examine small lesion. Another application of radiopharmaceuticals is in angiogenesis imaging. Integrin is one of the cell adhesion molecules which plays an important role in angiogenesis and metastasis. ^{99m}Tc RGD radiopharmaceutical probe has been introduced for early detection of angiogenesis and metastasis and has a potential application for antiangiogenic therapy [41, 42]. There is a special interest in the application of scintigraphy for following drug delivery using nanoparticles [43]. Application of nanocarriers such as carbon nanotubes, graphene oxides, liposomes, lipoplex, polymers, and so on to deliver bioactive nucleic acids, not only increase the biocompatibility and solubility, exhibit no immunogenic response as well.

Another application of scintigraphy is to evaluate multidrug resistance. In the absence of p-glycoprotein, radiopharmaceutical agents such as ^{99m}Tc MIBI able to translocate across the cell membrane and concentrate inside the cells due to its positive charge while the presence of p-glycoprotein makes ^{99m}Tc MIBI to pump out of the cells. So, radionucleotides can find out if the designed drug can effectively treat multidrug resistance or not.

2.5.2 Advantages and disadvantages

The advantages of using radiopharmaceutical imaging system list as follow:

- Identify skeletal problem
- Possibility of whole-body scanning
- Monitoring the treatment response

The disadvantages of radiopharmaceutical imaging are:

- Not widely available
- Relatively higher cost that X-ray and CT
- Low spatial resolution (5–10 mm)
- Radiation risk
- Usually required I.V. injection
- Disposal of radio-activity waste
- Slow image acquisition

2.6 PET and PET/CT

Positron emission tomography (PET) is a nuclear medicine imaging system applied to observe metabolic processes to diagnosis and follow up various diseases. PET and PET/CT have been used to study various tumors such as lymphoma, melanoma, lung carcinoma, pulmonary nodules, colorectal cancer, and breast cancer [44].

In general, MRI and CT are based on morphological changes for diagnosis and follow-up malignances while PET detects abnormal metabolic activity in target as yet doesn't undergo anatomical changes. Also, PET is able to monitor patients subjected to chemotherapy or tumor resection whom having difficulty to appearance using MRI or CT imaging systems.

Although PET has considerable advantages as an imaging system, it is hard to accurately find out the area of abnormal activity due to the lack of anatomical structure data. Therefore, great attempts have been done to couple functional information of PET with anatomical information. Beyer et al. introduced PET/CT scanner to coregister structural and functional images without moving the patient [45].

PET imaging is based on the detection of γ photons released from radionuclides during the annihilation of positron (β) with electron (Fig. 2.5). The photons possess energy around 0.511 MeV and are detected by scintillation crystals. Scintillation crystals are mainly made up of gadolinium silicate, lutetium oxyorthosilicate, and bismuth germinate. The value of 512 keV is the energy of the mass of an electron based on the law of energy conservation.

The main sources of positron-emitting radioisotopes are ^{11}C, ^{64}Cu, ^{18}F, ^{15}O, ^{124}I, ^{86}Y, ^{13}N, and ^{68}Ga. ^{18}F is one of the most widely studied radioisotopes due to the long half-life (around 109.8 min). The specific feature of ^{11}C is that mainly the tracers based on ^{11}C are endogenous substances such as fatty acid acetate, hydyroxyephedrine and amino acids methionine while other radioisotopes are analogous of endogenous component. The main drawback of using ^{11}C is related to its short half-life (around 20.4 min). However, it may be a good candidate for developing cost-effective cyclotrons [46]. Recently, drug-based tracers have absorbed great deal of attention such as [N-methyl-^{11}C]Vorozole and ^{18}F-RGD peptide.

2.6.1 Emission detection in PET

As already mentioned, a positron (also called positive electron) annihilates with electron causing the release of two photons moving in opposite directions having the energy of 511 keV. The photons then will be detected using scintillation crystals connected to photomultiplier tubes. Since the radionuclides typically use in PET imaging emit photons with higher energy than the radioisotopes use in nuclear medicine, utilizing detectors with higher stopping power is needed. Crystals which are mainly used for PET imaging have high atomic numbers and densities. The efficiency of coincidence detection for 25-mm bismuth germinate and lutetium oxyorthosilicate is around 80%. It should be noted that the absorption efficiency of bismuth germinate crystal is higher than lutetium oxyorthosilicate due to its higher atomic number, while lutetium oxyorthosilicate crystals emit higher light than bismuth germinate. Gadolinium silicate possess a lower atomic number than lutetium oxyorthosilicate and bismuth germinate. Gadolinium silicate crystals can detect photon energy in the wide range and emit slightly higher light than bismuth germinate.

The crystals are placed into block detectors. Block detectors compose of some channels fill with opaque materials and contain many small crystals. The advantages of using block detector over single crystal is that each small crystal acts as an independent system with lower detector dead time.

Due to the kinetic energy, proton travel a small distance before annihilation called mean positron range (as shown with "a" in Fig. 2.6). Based on the density of the environment, mean positron range is different, for example, it is smaller for bon

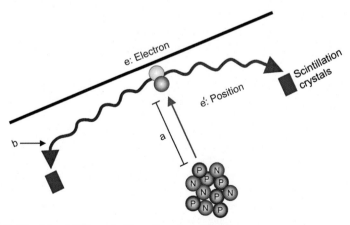

FIGURE 2.6 Principle of PET production. *P, proton*; *N*, neutron.

than air or lung. The change in positron position and the site of annihilation causes positron range blurring, decreasing the spatial resolution of PET systems. Another reason of limiting special resolution is related to the angle between annihilation photons. Positron and electron annihilate during motion which cause the created angle between emitted photons are not precisely 180°C (as shown with "b" in Fig. 2.5). This random variation in the predicted angle of 180°C lead to the additional decrease in spatial resolution called angle blurring [44]. These errors can be magnified based on the position of the detectors and the scanners' diameter.

It should be noted that the scattering of photons before detecting by the detector may cause additional decrease of image resolution.

2.6.2 Advantages and disadvantages

The advantages of PET/CT imaging technic are:

- Apply for several tumors
- Preforming whole-body imaging which will be helpful to detect metastasis tumor
- High resolution with perfect attenuation, resolution, and scattering correction
- Based on metabolic activity it can have high sensitivity
- Calcium scarring
- Having high accuracy due to the combined with CT

The disadvantages of using PET/CT imaging technic are explained as follow:

- Nonspecific for tumor type
- Costly
- Not applicable for infected or inflammation region
- Image artifacts may occur
- Inadequate information for surgical planning
- Shielding for CT

After introducing the effect of nonbiological systems in cancer diagnosis, in the following chapter, the effect of biological components will be reviewed.

References

[1] J.C. De La Vega, U.O. Häfeli, Utilization of nanoparticles as X-ray contrast agents for diagnostic imaging applications, Contrast Media Mol. Imaging 10 (2) (2015) 81–95.

[2] J. Wallyn, N. Anton, S. Akram, T.F. Vandamme, Biomedical imaging: principles, technologies, clinical aspects, contrast agents, limitations and future trends in nanomedicines, Pharm. Res. 36 (6) (2019) 78.

[3] M.A. Hahn, A.K. Singh, P. Sharma, S.C. Brown, B.M. Moudgil, Nanoparticles as contrast agents for in-vivo bioimaging: current status and future perspectives, Anal. Bioanal. Chem. 399 (1) (2011) 3–27.

[4] M.S. Kandanapitiye, M. Gao, J. Molter, C.A. Flask, S.D. Huang, Synthesis, characterization, and X-ray attenuation properties of ultrasmall BiOI nanoparticles: toward renal clearable particulate CT contrast agents, Inorg. Chem. 53 (19) (2014) 10189–10194.

[5] C. Briguori, D. Tavano, A. Colombo, Contrast agent-associated nephrotoxicity, Progr. Cardiovasc. Dis. 45 (6) (2003) 493–503.

[6] Widmark, J. M. (2007, October). Imaging-related medications: a class overview. In Baylor University Medical Center Proceedings (Vol. 20, No. 4, pp. 408-417). Taylor & Francis.

[7] J.M. Idée, B. Guiu, Use of Lipiodol as a drug-delivery system for transcatheter arterial chemoembolization of hepatocellular carcinoma: a review, Crit. Rev. Oncol. /Hematol. 88 (3) (2013) 530–549.

[8] C.E. Suckow, D.B. Stout, MicroCT liver contrast agent enhancement over time, dose, and mouse strain, Mol. Imaging Biol. 10 (2) (2008) 114–120.

[9] D.P. Cormode, T. Skajaa, Z.A. Fayad, W.J. Mulder, Nanotechnology in medical imaging: probe design and applications, Arteriscl. Throm. Vasc. Biol. 29 (7) (2009) 992–1000.

[10] J.D. O'Sullivan, J. Behnsen, T. Starborg, A.S. MacDonald, A.T. Phythian-Adams, K.J. Else, et al. X-ray micro-computed tomography (μCT): an emerging opportunity in parasite imaging, Parasitology 145 (7) (2018) 848–854.

[11] E.K. Oikonomou, M. Marwan, M.Y. Desai, J. Mancio, A. Alashi, E.H. Centeno, et al. Non-invasive detection of coronary inflammation using computed tomography and prediction of residual cardiovascular risk (the CRISP CT study): a post-hoc analysis of prospective outcome data, Lancet 392 (10151) (2018) 929–939.

[12] R. Tang, M. Saksena, S.B. Coopey, L. Fernandez, J.M. Buckley, L. Lei, et al. Intraoperative micro-computed tomography (micro-CT): a novel method for determination of primary tumour dimensions in breast cancer specimens, Br. J. Radiol. 89 (1058) (2016) 20150581.

[13] D.J. Brenner, E.J. Hall, Computed tomography—an increasing source of radiation exposure, N. Engl. J. Med. 357 (22) (2007) 2277–2284.

[14] C. Sun, J.S. Lee, M. Zhang, Magnetic nanoparticles in MR imaging and drug delivery, Adv. Drug Deliv. Rev. 60 (11) (2008) 1252–1265.

[15] J.M. Tognarelli, M. Dawood, M.I. Shariff, V.P. Grover, M.M. Crossey, I.J. Cox, et al. Magnetic resonance spectroscopy: principles and techniques: lessons for clinicians, J. Clin. Exp. Hepatol. 5 (4) (2015) 320–328.

[16] H.B. Na, I.C. Song, T. Hyeon, Inorganic nanoparticles for MRI contrast agents, Adv. Mater. 21 (21) (2009) 2133–2148.

[17] S. Mornet, S. Vasseur, F. Grasset, E. Duguet, Magnetic nanoparticle design for medical diagnosis and therapy, J. Mater. Chem. 14 (14) (2004) 2161–2175.

[18] G.J. Strijkers, M. Mulder, J. Willem, F. Van Tilborg, A. Geralda, K. Nicolay, MRI contrast agents: current status and future perspectives, Anti-Cancer Agents Med. Chem. 7 (3) (2007) 291–305.

[19] W. Chen, D.P. Cormode, Z.A. Fayad, W.J. Mulder, Nanoparticles as magnetic resonance imaging contrast agents for vascular and cardiac diseases, Wiley Interdiscipl. Rev.: Nanomed. Nanobiotechnol. 3 (2) (2011) 146–161.

[20] K.K. Shung, Diagnostic ultrasound: past, present, and future, J. Med. Biol. Eng. 31 (6) (2011) 371–374.

[21] A.J. Coleman, J.E. Saunders, A review of the physical properties and biological effects of the high amplitude acoustic fields used in extracorporeal lithotripsy, Ultrasonics 31 (2) (1993) 75–89.

[22] O. Al-Bataineh, J. Jenne, P. Huber, Clinical and future applications of high intensity focused ultrasound in cancer, Cancer Treat. Rev. 38 (5) (2012) 346–353.

[23] F. Kiessling, S. Fokong, J. Bzyl, W. Lederle, M. Palmowski, T. Lammers, Recent advances in molecular, multimodal and theranostic ultrasound imaging, Adv. Drug Deliv. Rev. 72 (2014) 15–27.

[24] D. Bokor, J.B. Chambers, P.J. Rees, T.G. Mant, F. Luzzani, A. Spinazzi, Clinical safety of SonoVue™, a new contrast agent for ultrasound imaging, in healthy volunteers and in patients with chronic obstructive pulmonary disease, Investig. Radiol. 36 (2) (2001) 104–109.

[25] S. Fokong, M. Siepmann, Z. Liu, G. Schmitz, F. Kiessling, J. Gätjens, Advanced characterization and refinement of poly N-butyl cyanoacrylate microbubbles for ultrasound imaging, Ultrasound Med. Biol. 37 (10) (2011) 1622–1634.

[26] G. Korpanty, P.A. Grayburn, R.V. Shohet, R.A. Brekken, Targeting vascular endothelium with avidin microbubbles, Ultrasound Med. Biol. 31 (9) (2005) 1279–1283.

[27] S.R. Sirsi, M.A. Borden, Microbubble compositions, properties and biomedical applications, Bubble Sci. Eng. Technol. 1 (1–2) (2009) 3–17.

[28] M. Postema, G. Schmitz, Ultrasonic bubbles in medicine: influence of the shell, Ultrason. Sonochem. 14 (4) (2007) 438–444.

[29] W.T. Joines, Y. Zhang, C. Li, R.L. Jirtle, The measured electrical properties of normal and malignant human tissues from 50 to 900 MHz, Med. Phys. 21 (4) (1994) 547–550.

[30] Y. Liu, H. Wang, Nanomedicine: nanotechnology tackles tumours, Nat. Nanotechnol. 2 (1) (2007) 20.

[31] M. Xu, L.V. Wang, Photoacoustic imaging in biomedicine, Rev. Sci. Instrum. 77 (4) (2006) 041101.

[32] L.A. Geddes, L.E. Baker, The specific resistance of biological material—a compendium of data for the biomedical engineer and physiologist, Med. Biol. Eng. 5 (3) (1967) 271–293.

[33] R.J. Halter, A. Hartov, K.D. Paulsen, A broadband high-frequency electrical impedance tomography system for breast imaging, IEEE Trans. Biomed. Eng. 55 (2) (2008) 650–659.

[34] L. Fass, Imaging and cancer: a review, Mol. Oncol. 2 (2) (2008) 115–152.

[35] A.P. Bagshaw, A.D. Liston, R.H. Bayford, A. Tizzard, A.P. Gibson, A.T. Tidswell, et al. Electrical impedance tomography of human brain function using reconstruction algorithms based on the finite element method, NeuroImage 20 (2) (2003) 752–764.

[36] Krebs, S., & Dunphy, M. (2017). Role of nuclear medicine in diagnosis and management of hepatopancreatobiliary disease. In Blumgart's Surgery of the Liver, Biliary Tract and Pancreas, 2-Volume Set (pp. 285-315). Content Repository Only!.

[37] Fogelman, I., Cook, G., Israel, O., & Van der Wall, H. (2005, April). Positron emission tomography and bone metastases. In Seminars in nuclear medicine (Vol. 35, No. 2, pp. 135-142). WB Saunders.

[38] G. Mariani, L. Moresco, G. Viale, G. Villa, M. Bagnasco, G. Canavese, et al. Radioguided sentinel lymph node biopsy in breast cancer surgery, J. Nucl. Med. 42 (8) (2001) 1198–1215.

[39] L.R. Coover, G. Caravaglia, P. Kuhn, Scintimammography with dedicated breast camera detects and localizes occult carcinoma, J. Nucl. Med. 45 (4) (2004) 553–558.

[40] M.K. O'connor, S.W. Phillips, C.B. Hruska, D.J. Rhodes, D.A. Collins, Molecular breast imaging: advantages and limitations of a scintimammographic technique in patients with small breast tumors, Breast J. 13 (1) (2007) 3–11.

[41] M. Fani, D. Psimadas, C. Zikos, S. Xanthopoulos, G.K. Loudos, P. Bouziotis, A.D. Varvarigou, Comparative evaluation of linear and cyclic ^{99m}Tc-RGD peptides for targeting of integrins in tumor angiogenesis, Anticancer Res. 26 (1A) (2006) 431–434.

[42] K.H. Jung, K.H. Lee, J.Y. Paik, B.H. Ko, J.S. Bae, B.C. Lee, et al. Favorable biokinetic and tumor-targeting properties of 99mTc-labeled glucosamino RGD and effect of paclitaxel therapy, J. Nucl. Med. 47 (12) (2006) 2000–2007.

[43] N. Sharma, S. Tyagi, S.K. Gupta, G.T. Kulkarni, A. Bhatnagar, N. Kumar, Development and gamma-scintigraphy study of Hibiscus rosasinensis polysaccharide-based microspheres for nasal drug delivery, Drug Dev. Ind. Pharm. 42 (11) (2016) 1763–1771.

[44] V. Kapoor, B.M. McCook, F.S. Torok, An introduction to PET-CT imaging, Radiographics 24 (2) (2004) 523–543.

[45] T. Beyer, D.W. Townsend, T. Brun, P.E. Kinahan, M. Charron, R. Roddy, et al. A combined PET/CT scanner for clinical oncology, J. Nucl. Med. 41 (8) (2000) 1369–1379.

[46] B. Långström, O. Itsenko, O. Rahman, [11C] Carbon monoxide, a versatile and useful precursor in labelling chemistry for PET-ligand development, J. Label. Compd. Radiopharm. 50 (9–10) (2007) 794–810.

Immune assay assisted cancer diagnostic

3.1 Polyclonal antibody-based immune assays

Cancer is the second leading cause of death in the world. Early detection represents one of the most promising approaches to reducing the growing cancer difficulty. The challenge consists of detecting tumors at early stages to make possible therapeutic treatment before progression occurs. This aim is mostly important in high-risk populations in which the occurrence of this disease is significantly increased. For early diagnosis to be a beneficial and practical approach, screening methods must satisfy five basic requirements:

1. The method must show a high degree of precision with an acceptable cut-off level defined and agreed.
2. Diagnosis should be possible at stages where disease is medicable.
3. The technique should enable discrimination between aggressive lesions requiring treatment from harmless lesions, avoiding the trouble of overdiagnosis.
4. The technique should be low-cost and well accepted by the target population [1].
5. The technique should be reproducible and correctly calibrated to be applicable [1,2].

Bio-Engineering Approaches to Cancer Diagnosis and Treatment. http://dx.doi.org/10.1016/B978-0-12-817809-6.00003-0

Table 3.1   Biomarkers in cancer screening studies (cancer site).

Tumor marker	Cancer site	Sample matrix
HE4	Ovary	Blood
CEA	Ovary, kidney, colon	Blood
CA-125	Breast, liver, lung	Blood
HER2	Breast	
AFP	Stomach	Blood
CA 15-3	Breast, lung, ovary	Blood
β-hCG	Germ cell of ovary	Blood, urine

There are several ways to reach a cancer diagnosis, including ultrasound, magnetic resonance imaging (MRI), X-ray computed tomography (CT), biopsy, and blood tests for cancer-related markers. Tumor markers are the proteins that have presented in circulatory system and their elevated levels can indicate the stage of the cancer. Some of the biomarkers which have been used in recent cancer screening studies are listed in Table 3.1.

Immunoassay has been known as one of the uppermost analytical techniques and is widely used in clinical diagnoses and biochemical studies because of its particularly high selectivity and sensitivity [3].

One of the immune system's principal functions is the production of circulating molecules called antibodies or immunoglobulins (Igs). Each Ig has comprised of a specific antigenic determinate or epitope (Fig. 3.1). Although an antigen may have many different epitopes that react with several different Igs, a unique Ig joins with only one epitope and this epitope (and the antibody's combining site for it) has a size of about five to seven amino acids.

Anticancer antibodies (polyclonal and monoclonal) have been used since last time in the diagnosis and therapy of cancer, with major applications having been

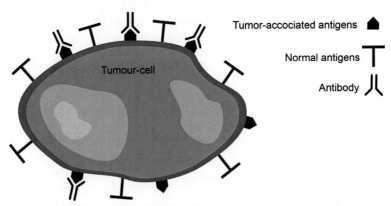

FIGURE 3.1 Detection of tumor-associated antigens expressed in cancer cells using antibody.

Table 3.2 Features of polyclonal and monoclonal antibodies.

Characteristic	Polyclonal antibody	Monoclonal antibody
Antibody purification	May/may not	Not essential
Booster dose	Required	Not required
Chemically defined	Not well	Well
Clonality	Several	Single
Cross-reactivity	High to low	Low to nil
Epitope detection	Multiple	Single
Homogeneity	NO	Yes
Payload (conjugation)	Difficult	Easy
Specificity	Low	High
Variability	High	Low

shown in the immunohistochemistry and immunoassay of tumor-associated antigen (TAA) mark. Researchers have used different polyclonal antibodies (PoAbs) for cancer diagnosis, such as purified polyclonal antibody against the human homologs of CD44 variant exon sequence to investigate the presence of CD44 on primary invasive breast tumors and lymph node metastases or 47 kDa PoAbs for detection of bladder cancer [4].

In 2018, Prasetya et al. have concluded that urinary detection of bladder cancer cells with 47 kDa PoAbs has higher sensitivity values and negative predictive values (100%) than urine cytology in patients with hematuria. The specificity of immunocytochemistry using 47 kDa PoAbs was 36.36%. The positive predictive and negative predictive values of immunocytochemistry using 47 kDa PoAbs were 22.22% and 100.00%, respectively. This technique can be used as an early detection method for bladder cancer [5].

Both of these antibodies have similar structures and functions, but PoAb and monoclonal antibody (MoAb) are different from each other on the basis of their first principle, production, and specificity. The difference between these two antibodies is based on the clonality of the cells that produce them. MAbs are produced by a single clone, while PoAbs are produced by numerous clones together. However, problems such as lack of adequate specify have limited the use of these antibodies for diagnoses. Some properties of PoAb and MAb are listed in Table 3.2 for a better understanding [6].

3.2 Monoclonal antibody-based immune assays

MoAbs have long been an appropriate tool in basic research due to their high specificity and affinity for tumor-specific antigens [7].

MoAbs should be highly sensitive for the TAA and have negligible cross-reactivity with normal tissue. Antigen properties affecting the precision of MoAb using in vivo include a cellular versus a subcellular location, antigenic stability, binding site availability, degree of homology with other molecules, and degree of tumor antigen expression throughout the tumor. Additional variables include type of tumor, size, viability, position and vascularity of the tumor, route of management, and the procedure and scanning instrument have used.

Furthermore, antibodies which have been injected into the bloodstream pass through a number of compartment, including vascular and extravascular spaces (organs, tissues, and body fluids), and there are many obstacles to the delivery of an injected MoAb. It has demonstrated that only a small amount of the injected Ig (approximately between 0.0007% and 0.01% of the injected dose per gram of tissue) targets the tumor. For instance, the binding of MoAbs to circulating antigen (secreted from the tumor) or to antigen present on normal tissue can deflect MoAbs from the tumor. At the tumor site, poor vascular penetration and vascularization can also decrease MoAb availability and prevent binding to tumors. At the cellular level, antigen modulation, unavailability, and low density can all affect the amount of MoAb bound to tumor. The site of tumor(s) in the body may affect the technique of MoAb management. Intravenous injection is the most popular technique, but when the malignancy is limited to a region or body cavity, intraregional or intracavitary usage, such as intraperitoneal or intrathecal injection, has demonstrated effective for imaging.

Effectiveness of both immune scintigraphy and immunotherapy depends on the portion of antibody taken up in the tumor. This depends on penetrability of MoAbs in the tumor cells. Tumors are characterized by heterogeneous and high interstitial fluid pressure with high viscosity of tumor blood supply. MoAbs have to penetrate through this pressure gradient which finally depends on the molecular size of MoAbs. Thus, the large size of MoAbs, because of the presence of Fc region, is an inconvenience in cases where tumor penetration is difficult due to heterogeneity. Whole antibodies have a multinodular nature with each domain having a particular function, that is, Fab region for antigen binding and Fc domain for effector functions. Interaction of the neonatal Fc receptor (FcRn) and Fc domain of antibody aids to increase its biological half-life. However, recently the concept of engineered antibody fragments and single-domain antibody (sdAb) known as mini bodies has been developed by through selection of suitable molecular domains of MoAbs in efforts to control in vivo valence, affinity, avidity, and tissue [8]. Antibody fragments are often used in cases such as radiolabeling [9] and the combination of imaging and immune diagnostic and will be discussed later in this chapter.

The most important function of anticancer MoAbs has been for the measurement of circulating TAAs, such as CEA, AFP, hCG, PSA, CA125, in body fluid immunoassays, and so forth.

For example, RP215 is a MoAb produced in mice immunized against OC-3-VGH ovarian cancer extract and it was shown that it mainly reacts with protein bands of 55 kDa, when Western Blot assays were carried out either with cancer cell extract or with affinity-purified CA215 [10]. RP215 is specific to carbohydrate-associated epitope of CA215, consists mainly of cancerous antigen receptors. So RP215 can be used for quantitation of CA215 in serum specimens of patients with cancer. RP215 has been used as the unique probe for cancer diagnosis.

A comparative research list of the positive rates of serum CA215 levels in different types of cancer including: lung (52%), colon (44%), ovary (59%), breast (38%), pancreas (51%) esophagus (61%), stomach (60%), kidney (38%), and lymphoma (83%) [11]. Immunohistochemically tissue staining studies with RP215 as the immune probe for cancerous antigen receptors also have indicated widespread expressions of RP215-epitope among various cancer tissues. Among these, the results were: ovary ($n = 87$, 64.4%), cervix ($n = 51$, 84.3%), endometrium ($n = 36$, 77.8%), stomach ($n = 93$, 49.5%), colon ($n = 87$, 43.6%), esophagus ($n = 56$, 75.7%), lung ($n = 58$, 31%), and breast ($n = 59$, 32.2%). In contrast, low percentages of positive tissue staining were observed for the cancer of liver ($n = 60$, 3.5%) and prostate ($n = 22$, 10%). Based on the results, RP215 could be a unique probe for cancer diagnosis [12,13].

In recent years, evidence of circulating autoantibodies in the sera of patients with cancer have created opportunities for utilizing the immune system as a source of cancer biomarkers. Autoantibodies are used as serum biomarkers which show highly appealing features. The persistence and stability of autoantibodies in the serum of patients is a good point over other potential markers that have been currently utilized. Autoantibodies are available in the sera before of TAAs can be detected (if they are). They correspond to an effective biological amplification of the presence of TAAs, and are released in the serum prior to initial clinical symptoms. Moreover, antibodies are highly stable in serum samples and are not subjected to proteolysis like other polypeptides, making sample handling much easier. They have shown a long lifetime (half-life between 7 and 30 days, depending on the type of Ig) in the blood and may persist as long as the corresponding autoantigen creates a specific humoral response. A few researches have attempted to identify panels of autoantibodies for early diagnosis of breast cancer. The results were favorable, in some patients sensitivities above 70% [14,15].

A summary of the recent studies on the use of autoantibody for the diagnosis of cancers is given in Table 3.3.

Table 3.3 Autoantibodies associated with cancer detection.

TA autoantibody marker	Cancer type	Method	Sensitivity	Specificity	Study year	Ref.
P53	Lung	Immunoassay	76%	73.2%	2016	[16]
P16, c-myc, ANAX-1	Breast	Elisa	75%	90%	2017	[17]
P53,CA19-9	Colorectal	Elisa	47.1%	None	2016	[18]
Six peptide clones	Lung	Immunohistochemistry	95.6%	95.6%	2010	[19]

There are several methods for identification of TAA and autoantibody including SEREX, phage display, protein microarray, SERPA, and Mapping.

Serological analysis of tumor antigens by recombinant cDNA expression cloning (SEREX):

In 1995, Serological analysis of tumor antigens by recombinant cDNA expression cloning (SEREX) was developed. SEREX involves the recognition of TAAs by screening patient's sera against a cDNA expression library acquired from the autologous tumor tissues. Afterward, a large number of TAAs associated with many cancer types have been detected using this technique. The usage of SEREX has facilitated the detection of TAAs as potential cancer biomarkers. This method in various types of cancers, including lung, liver, breast, prostate, ovarian, renal, head and neck, esophageal, leukemia, and melanoma have been used. There are, however, some restrictions to the SEREX approach. First, TAAs detected by SEREX are mostly linear epitopes. Second, there is a bias toward antigens that are overexpressed in the tumor tissues used to make cDNA libraries. Thus, overexpression of the antigens is often causes by their immunogenicity identified by SEREX. However, TAAs that are of low quantities are missed by SEREX. Third, SEREX is time-consuming, labor-intensive and cannot be automated. Thus, this method is not useful for analyzing a large number of patient serum samples. Finally, posttranslational modifications (PTMs) cannot be recognized by SEREX.

3.2.1 Phage display

In the phage display technique, a cDNA phage display library is created using a tumor tissue or cancer cell line. Peptides from the tumor or cell line are expressed as fusions with phage proteins and are displayed on the host surface. This aspect of the method allows inexpensive and labor-effective screening during biopanning. This technique also eliminates proteins that cannot be displayed on the surface of the phage species. Although this technique is effective for throughput than SEREX, it still cannot detect antigens with PTMs (e.g., glycosylated cancer antigens).

3.2.2 Serological proteome analysis (SERPA)

Another method used in the detection of TAAs is the proteomics-based technique termed SERPA or proteomics which uses a combination of 2D electrophoresis, western blotting, and mass spectrometry (MS). Proteins from tumor tissues or cell lines are obtained by 2D electrophoresis, and have transferred onto membranes by electroblotting and analyzed with sera from healthy people or patients with cancer. These include biases to abundant proteins and barriers in resolving certain types of proteins and in complexity producing reproducible 2D gels. Because of the method that western blots are prepared, only linear epitopes can be recognized [20]. SERPA has been used in the study of many cancers, such as neuroblastoma, lung carcinoma, breast carcinoma, renal cell carcinoma, HCC, and ovarian cancer. For example, the use of SERPA has recognized calreticulin and DEAD-box protein 48 (DDX48) in pancreatic cancer [21].

3.2.3   Multiple affinity protein profiling (mapping)

Mapping involves 2D immunoaffinity chromatography, digestion of the isolated proteins by enzyme, and identification of TAAs by consecutive nano liquid chromatography-mass spectrometry (nano LC-MS /MS), may enable the identification of these unknown antibodies of patient. In the first step of the chromatography, nonspecific TAAs in a cancer cell line or tumor tissue lysate such as colon cancer cells bind to IgG gotten from healthy controls in the immunoaffinity column and then are deleted from the lysate. The "flow-through fraction" of the lysate is then put to the 2D immunoaffinity column that comprises IgG from patients with cancer. TAAs that bind simultaneously are likely to be cancer-specific and are removed for enzymatic digestion and detection by consecutive MS [20].

Finally, antibodies are biochemically well-known molecules, and many available reagents and techniques are available for their detection, simplifying assay development.

3.2.4   **Proteomic microarray**

One of the necessities needed to improve the survival rate of patients with cancer is early diagnosis. Gene expression is a key to cellular processes, like stem cell maintenance, cell cycle, and cellular differentiation, as well as response to environmental changes. Any epigenetic alteration in control of gene expression could lead to the many different diseases such as the formation of tumors. DNA methylation was the first epigenetic mark known to be a crucial cause of cancer. The difference between tumor cell and normal cells is because of aberrant changes in the methylation pattern or chromatin modification which is occurred in cells. Therefore, by comparing malignant cells with normal cells, early detection of cancer would be feasible before any symptoms have appeared. The latest advances in research have shown that PTMs of histones (such as phosphorylation, acetylation, or methylation) were also found to be involved in tumorigenesis [22,23]. Interestingly, more than 200 different PTMs are identified in proteins. The PTM of the proteome is a dynamic process that adjusts the cell signaling process. PTM has the inherent characteristics that have the ability to produce changes in macromolecules that potentially affect cancer activation, progression, and therapeutic response. This feature of PTM has been used in cancer diagnosis.

Detection of PTM and biomarkers is an important method for cancer diagnosis and in this way proteomic technology has played an important role in biomarker discovery and early detection. Clinical samples of proteins can be analyzed through procedures, such as MS, two-dimensional polyacrylamide gel electrophoresis (2D PAGE), and protein arrays in order to help compare the protein variability between samples. Table 3.4 describes the advantages and disadvantages of proteomic technology which has been investigated for cancer diagnosis [24,25].

Each proteomic technique has its own advantages or disadvantages and based on researcher's requirement or its own feature are used in different fields. Another important feature of a useful proteomic techniques is its noninvasive feature, which

Table 3.4 Advantages and disadvantages of proteomic technology.

Proteomic technology	Advantages	Disadvantages
ELISA	Very robust, it has the highest sensitivity, good technology for specific and single marker detection. It could be used for marker detection in body fluid or tissue	Requires well-characterized antibody for detection and extensive validation. Not amenable to direct discovery
2D-PAGE	It has a direct identification marker. Reproducible and more quantities combined with fluorescent dyes	It has a low sensitivity and efficiency. Can't be used as a direct means for early biomarker detection
MudPIT	Has high sensitivity, can be used for detection and identification of potential biomarkers, have a vast coverage in biomarker detection	It needs identification for pattern diagnosis
Proteomic pattern diagnosis	It doesn't need any protein IDs	It has medium sensitivity and weight limit for biomarkers. It can't be used for direct identification markers
Protein microarrays	It has high sensitivity with high efficiency. Format flexible and it could be used in any situation. It has the ability to detect PTMs	It's limited by antibody sensitivity, required enough knowledge to be used

means that it can detect markers with high sensitivity from body fluids such as serum, plasma, and urine. Proteomic technique is not sufficient for high accuracy diagnosis. In a recent study, some gene alternation like somatic mutation is characterized in patient with breast cancer, however, it is poorly understood with proteomic technique. Mertines et al. have shown that usage of MS analysis with proteomic array techniques would increase the region of discovery in cancer. They have selected 125 breast cancer samples and found 77 high-quality data. They also have identified G-proteins what wasn't easily detected by miRNA techniques [26]. This highlights that this combined approach can be a worthy method for accurate identification. Recently studies have shown that functional proteome positively complements genomic and transcriptomic data, and this approach has the ability to detect new biomarkers. In a pan-cancer study, researchers have found a relationship between HER2 variation, mRNA expression, and protein expression levels in different stages of cancer. The detail molecular landscape which has been obtained by this approach could not be concluded by just analyzing either DNA or RNA alone. This case report has supported the importance of combination of both techniques in biomarker detection [22–26].

Among proteomic technology, proteomic array is known as the best one, because of all positive features. The proteomic array has divided into three techniques: (1)

Analytical protein array that consists of antibodies or lectins to detect and/or quantify a large number of proteins present in a biological sample. (2) Functional protein arrays have used in studying the biochemistry properties of proteins, such as protein binding activities and enzyme-substrate relationships. (3) Reverse-phase protein arrays, mostly have used to detect signaling pathways. Antibody-based microarray is currently used in cancer diagnosis as it is noninvasive and has a high sensitivity for detection of biomarkers and PTMs [25].

3.3   Antibody-based microarray

The microarray technique was first has introduced in the 1980s. The main aim of this technique in cancer diagnosis was based on genome detection. This technique is separated into four branches: microarray DNA, microarray protein, antibody-based microarray, and microarray carbohydrate [22,24,25]. This technique is based on the binding or hybridization of the sample with the template. Over time, the sensitivity of the microarray has improved and its use has changed. Antibody-based microarray is one of the microarray techniques that is used today for the detection of specific biomarkers with high sensitivity. It has the benefit of rapid diagnostic speed and cost-efficiency. In addition, microarray is a noninvasive procedure, and it is highlighted in a study by Karen et al., which has demonstrated higher levels of serum protein in patient plasma of cancer patients compared to control samples [27].

A drawback of protein biomarkers is a lack of specificity, with raised biomarkers suggestive of more than one disease. Serum PSA is a protein that is elevated in some benign prostate diseases as well as prostate cancer, or interleukin 6 (IL-6), is also a protein biomarker which, if overexpressed in the serum of patients, could be known as inflammation of the prostate, lung, multiple myeloma, and renal cell cancer Thus, single cancer biomarkers are often not unique to a specific cancer. Antibody-based microarray has the ability to concurrently detect multiple specific biomarkers, and target-specific cancers using a panel of biomarkers [24,22].

In another study, further advantage of utilizing antibody microarray for early-stage diagnosis has been reported. Pancreatic cancer is an aggressive disease with poor prognosis, and most people die because of a delayed diagnosis. In the past, biomarkers such as CA125 or CA19-9 could not be detected until advanced pancreatic cancer stages; however, the high sensitivity technique which has acquired through antibody-based microarray has caused such markers can be used before any crucial progression occurs [28]. In an investigation of 148 patients with early-stage pancreatic cancer, chronic cancer, autoimmune pancreatitis (AIP), and control patients, Wingren et al. have identified a panel of 25 protein targets such as IL-2, IL-11, IL-12, TNF which distinguish pancreatic cancer from healthy controls. In pancreatic cancer, these 25 biomarkers have shown a high sensitivity in cancer diagnosis (AUC 86%) [28].

Moreover, microarray is also used in the diagnosis of prostate and breast cancer. These findings are useful to aid the development of novel and multivariate diagnostics methods. In a recent study, researchers have found three proteins of ERBB2, TNC, and ESR1 for pancreatic ductal adenocarcinoma (PDA) which increases the AUC from 0.86 to 0.97. Using this technique, CA19-9 was discovered to play an important role in early cancer diagnosis [22,24,28].

Antibody-based microarray is also used for understanding the mechanism of tumor progression. In bladder cancer, it is seen that 50%–70% of cancers recur after standard transurethral resection. With the help of the antibody microarray method, Srinivasan et al. have found 20 proteins that facilitated the tumor progression, leading to the prediction of recurrence with sensitivity of 80% and specificity of 100%. They have also found repression of the TGF-β signaling pathway in recurrent cancer [29]. They have reported that the signaling factors IFNG, TNF-α, and THBS1 were expressed less, and the abundance of the inhibitor MAPK3 (also known as ERK1) was higher, whilst SMAD2, SMAD3, and SMAD4 were again significantly underrepresented in these patients with bladder cancer. The data has indicated that TGF-β signaling pathway inhibitors may reduce bladder cancer recurrence [29].

Gastric cancer (GC) has among the worse survival rates of any solid tumor. Through using antibody microarrays, Puig-Costa et al. were able to detect numerous biomarkers for GC, including 120 cytokines, 43 antigenic factors, 41 growth factors, 40 inflammatory factors, and 10 metalloproteinases [30]. This technique was used because it has the ability to detect different proteins in a short span of time, it is cost-effective and highly sensitive. This has also helped to discover of some biomarkers such as ICAM-1 and angiotensin that indicate a high inflammatory response in gastric disease. It has been shown that monocyte chemotactic protein 1 (MCP-1) is associated with tumor development in GC, and can be used as a specific biomarker for GC patients. Antibody microarray analyses of GC has detected a 21-protein inflammatory protein-driven gastric cancer signature (INPROGAS) that accurately discriminates GC from noncancerous gastric mucosa, and may provide new leads for the analysis of cancer progression. In 2017, Quan et al. have added 39 additional markers, mostly consisting of cytokine, as it is understood that inflammation is a specific feature of GC. They have included biomarkers such as TGF-β, TNF, and mitogen-activated protein kinase (MAPK) signaling pathway, as a useful biomarker for early-stage cancer diagnosis [31].

In another study, Schwenk et al. have used suspension bead array technology for protein profile plasma in patients with prostate cancer and respective controls [32]. The results have indicated that carnosine dipeptidase 1 (CNDP1) plasma levels were decreased in patients with prostate cancer, and subsequently they have founded an association with aggressive prostate cancer. For increased sensitivity, a sandwich immunoassay has been developed to investigate these findings in 1214 patients, which has elucidated the association between decreased CNDP1 and lymph node metastasis. In another investigation, CNDP1 levels were suggested to have a metabolic role, as this investigation found reduced plasma levels of CNDP1 in cachexic patients [32]

Miller et al. have reported comparison between two microarray methods; antibody microarray and hydrogel bed. They have found that antibody-based microarray not only facilitates more rapid marker discovery but also enables the direct observation of the relationships between proteins. Thus from single data sets, one could examine a combination of multiple markers that may increase the statistical significance of a diagnosis [33].

In an investigation of small intestine neuroendocrine tumors, Darmanis et al. have used targeted bead arrays to investigate marker levels in plasma in two independent study sets (77 and 132 samples, respectively) for 124 unique proteins. The classification accuracy was up to 85%. They have used innovative biomarkers between the significant panel to investigate them. Among these candidates, IGFBP2 and IGF1 were indicated as new markers, thus the researchers have tested this through an ELISA assay and subsequently have confirmed this indication [34].

Within the last decade among proteomic array techniques, antibody-based microarray has been the main technique in development and detection of new biomarkers in cancer diagnosis. However, despite its advances and advantages (e.g., quality controls, specificity, functionality, and/or reproducibility), this technique is not sufficient alone and requires the input of other diagnostic techniques to complete validation.

3.3.1 PTM and its role in cancer diagnosis

One of the main reasons that antibody microarrays are a crucial technique in cancer diagnosis is their ability to detect PTMs. A histone modification is a PTM of histone proteins, which involves phosphorylation, acetylation, and methylation [35,36]. Phosphorylation, acetylation, and glycosylation activate signaling pathways and change normal cellular function. Recent studies have shown that PTMs have a crucial role in tumorigenesis and can provide useful information about the epigenetic regulation of cellular processes. For example, modifications in cell surface receptors such as receptor tyrosine kinases and G-protein coupled receptors can influence signaling pathways and contribute to tumorigenesis. Phosphorylation is the best-characterized modification in tumorigenesis. Until recent advances in antibody microarray technology, the study of PTMs and their role in cancer diagnosis and prognosis had been very limited.

Today, the role of phosphorylation in protein function is well characterized, with the role of glycosylation, ubiquitination, and acetylation in tumorigenesis being intensely investigated. During tumorigenesis, genetic alterations in signaling molecules lead to the overactivation cell surface receptors, which consequently affects downstream signaling pathways. For instance, membrane receptors, like HER2 and FGFR, and components of the intracellular signaling cascade such as the K-RAS and ERK kinases, could join and play abnormal signaling. PTM, like the phosphorylation of STY, causes aberrant alternation and thus cells face to altered signaling. Recent investigations have shown that genetic alterations, such as somatic or germline mutations, modulate the functional activity of protein kinases (including multiple receptor tyrosine kinases and phosphatases in the genome) which has a functional impact at the proteome level. In summary, genomic alterations due to mutations functionally overactivated or alter signaling pathways, which make cells susceptible to neoplastic growth.

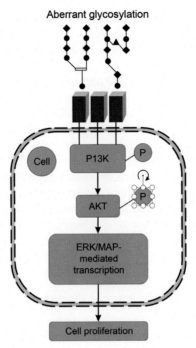

FIGURE 3.2 Mechanism of PTM.

Glycosylation is another important PTM involved in neoplastic transformation. Aberrant alterations in glycosylation patterns have been linked with tumor aggression and tumor microenvironment heterogeneity (Fig. 3.2). N-linked glycosylation is involved in a variety of cellular functions, such as cell-cell interactions, for example, metastasis and cancer progression. The discovery of an N-linked glycosylation site is important in the regulation and function of BRCA1 in breast cancer, as it is a potential breast cancer biomarker. In ovarian cancer, many membrane proteins have been found to be aberrantly glycosylated and modified, including CA125 and KLK6, which qualify as potential biomarkers for an early diagnosis. Further, MUC-4—a transmembrane protein—expression has been observed in PDA, and it is aberrantly glycosylated, and it is involved in cancer progression and neoplasm cancer aggression. Thus, MUC-4 can be a useful target in the development of novel therapeutic strategies for the treatment of pancreatic cancer. Additionally, a comparative proteomic analysis of three breast cancer cell lines (MCF-7, MDA-MB-453, and MDA-MB-468) has identified three N-linked glycosylated membrane proteins, namely galectin-3 binding protein, lysosome-associated membrane glycoprotein 1, and oxygen-regulated protein, respectively. Analyzing N-glycoproteins from the membranes of breast cancer cell lines have highlighted potential biomarkers for breast cancer diagnosis and thus potentially promises future therapy. Furthermore, genetic and epigenetic modifications on many glycogens are associated with malignant transformations. Through the recent advancements in proteomic technologies for cancer-cell glycemic, many tumor-associated glycoproteins and glycoproteomics have been utilized for diagnostic, prognostic, and therapeutic purposes. Furthermore, tumor-associated glycoantigens generate serum antibodies, which have potential applications as biomarkers for early breast cancer detection. The detection of aberrant glycosylated MUC1-specific autoantibodies correlates with colorectal cancer, which has the capability to predict cancer with 95% specificity. However, the low sensitivity of this marker could be used in conjunction with other markers, suggesting that a combination of antibody signatures may eventually enable a biomarker panel for the early detection of breast cancer.

Acetylation also plays an important role in the regulation of numerous Oncoproteins which are involved in tumorigenesis and cancer progression. Protein acetylation is involved in several processes, including cancer. Lysine N-acetylation precisely regulates the function of histone and nonhistone proteins, and especially, histone acetyltransferases (HATs) are dysregulated because of numerous genetic or epigenetic alterations. Normally, HATs act as tumor suppressors and facilitate normal cell growth, cell cycle, and help control of oncogenes. However, abnormal acetylation could activate malignant proteins and trigger tumorigenesis. Abnormal acetylation profiles have been used as diagnostic markers for early cancer detection. Moreover, acetylation also has potential as a prognostic biomarker to monitor cancer treatment. Further, epigenetic therapy, employing histone deacetylase inhibitors and acetylation modulators, has shown promising results in treating some forms of cancers. Some of the proteins involved in controlling N-acetylation and their targets are aberrantly regulated during tumorigenesis, and small molecule inhibitors such KAT, KDAC, and bromodomain are being tested as potential anti-cancer therapies to treat relapsed or refractory cutaneous T-cell lymphoma.

In conclusion, PTMs play an extremely important role in cancer activation by altering signaling pathways controlled by kinases. Thus, the phosphorylation of STY influences the kinases-phosphorylation network, which also alters responses to adjuvant therapy.

The glycosylation of membrane receptors such RTK and GPCR could also play an important role as a biomarker for disease diagnosis and in monitoring the effectiveness of neoadjuvant and adjuvant therapy [37].

By investigating of these techniques and familiarity with PTM, it is important to discover techniques that can detect PTMs. In proteomic techniques, earlier researches have been confined to single biomarkers, and several reports have characterized a single biomarker for the prediction of early cancer diagnosis and prognosis. However, it has been observed that single proteomic biomarkers lack the precision to accurately detect cancer and its clinical effectiveness is very limited. Consequently, the use of multigene expression signatures or a panel of proteomic signatures in a tumor sample that displays a distinct expression pattern and an enhanced diagnostic precision, are promising candidates as early-stage biomarkers.

3.4 Antibody-based immunosensors

Biosensor is an analytical device with the aim to detect chemical substances. Its works is based on the combination of the physicochemical detector and biological component. This biological component could be tissue, microorganism, enzymes, antibodies, antigens, nucleic acid which could interact and bind with specific analytic [38]. Immunosensors are kind of biosensors based on interactions between an antibody and an antigen on a transducer surface (Fig. 3.3). Antibodies and antigens can bind together with great strength, therefore they can be immobilized on a transducer, and detect each other. It also brings high sensitivity to these techniques too. Moreover, as each antibody or antigen can interact with their own antigen or antibody, therefore, immunosensor-based antibody has high selectivity too. Immunosensors based on its detection ways have divided into three parts: 1—electrochemical, 2—optical, 3—piezoelectric immunosensors (electrochemical quartz crystal microbalance). Immunosensors can be either direct or indirect. It means they have operated directly via the Ab/Ag interaction, or indirectly with a further label, such as an enzyme or fluorescent molecule, in order to detect whether a binding event has occurred.

Electrochemical immunosensors explore measurements of an electrical signal produced on an electrochemical transductor. Immunosensors based on electrochemical detection is used more in research because this technique is portable, simple and has an automated detection, meaning that it does not require handling by a person. Immunosensors have several parts, and each part is important and impacts development. An electrode surface is the first part. An electrode surface can be modified with gold nanoparticles, carbon nanotubes, nanofibers or any other conducting polymer. What is important in this part is using materials that increase conductivity, allowing for high sensitivity in detection. This conductive electrode has stayed on

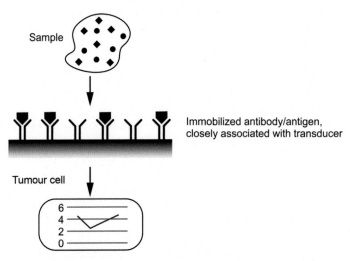

FIGURE 3.3 Schematic representation of antibody-based immounosensors.

immobilization platforms where bio receptors are set on. The second part is the bioreceptors, the antibody or aptamer that is on the surface where the modified electrodes exist. The system can be used after modification. The mechanism of this technique is based on the interactions between the immobilized bioreceptors and proteins inside the sample. And the reactions between them, like antibody-antigen or aptamer-antigen labeled with enzyme or an electroactive compound create measurable signals. In order to obtain good results from this technique, other factors such as pH, buffers, temperature, concentration of nanostructures, monomers and incubation times with aptamers, tumor markers, antibodies should be considered [38–40].

Optical immunosensors is another immunosensor technique that its sensor response to light modified upon binding of specific antigen. In this technique an optical sensor uses light for stimulation and then is able to detect alterations in the power of light as it passes through or refracts from a sampling system in relation to Ab/Ag binding. Optical immunosensors include surface plasmon resonance (SPR) based sensors, fiber-optic sensors (FOS) and various fluorescence-based sensors. Like electrochemical immunosensor these technique has some disadvantages that now using nanoparticles such as carbon nanotube has developed its efficiency it biomarker detection in body fluids.

Piezoelectric immunosensor operation is based on the conversion of physical or mechanical changes into electrical energy and vice versa. The commonest piezoelectric sensor is the quartz crystal microbalance (QCM). With this sensor when antibody or antigen is immobilized on the crystal surface, their bonding can be measured electrically. Development in this technique has increased its popularity. In 2017, Su et al. have developed a new biosensor based on the piezoelectric ceramic to detect cancer biomarkers. It uses two piezoelectric resonators in order to decrease the environmental influence such as temperature. This device has shown high sensitivity in detection of cancer markers prostate-specific antigen (PSA) and α-fetoprotein (AFP) and also it's detection is faster than using only one piezoelectric ceramic [41].

As it is said, biosensor immunoassay is an attractive assay for biomarker detection because it has a high sensitivity. The performance of any immunosensor is largely dependent on the type of antibody, and the associated antibody immobilization approach used to determine the sensor. Antibodies play a crucial role in delineating the sensitivity and specificity of an immunosensor. A necessary factor in immunosensor sensitivity is the immobilization strategy used with the antibody or biorecognition ligand [42]. Several studies have demonstrated the enhancement in antigen-binding activity by oriented antibody immobilization on a sensor surface. By production of recombinant antibody fragments, because of their small size, they are more stable and can be easily genetically changed to have highly oriented immobilization on the sensor surface. Recombinant antibodies show significant additional promise for the generation of antibody-based sensors with many novel applications in cancer diagnostics. In 2016 Spain et al. have introduced a sensitive electrochemical immunosensor that use recombinant scVF as a receptor on electrocatalytic platinum nanoparticle in order to detect PSA in serum sample. This device has detected a high concentration of PSA without need of any other techniques such as PCR or

NASBA, amplification [43]. Arkan et al. have investigated the detection of HER2 in serum samples of patients with breast cancer [44]. They have used gold as a nanoparticle due to its increased sensitivity, and discovered that by increasing the concentration of the HER2 antigen, detection is increased in serum samples of patients with breast cancer.

Electrochemical and optical biosensors and techniques are particularly attractive for biomarker detection because of their high sensitivity, relatively easy fabricating, easy operating procedures, and the potential to be miniaturized. Another technique that can improve the sensitivity of immounosensors is carbon nanomaterials because of their excellent electric and mechanical properties. The combination of carbon nanomaterials into biosensor platforms is now popular in biosensor design. The most widely used carbon nanomaterials to date are nanotubes [single-walled (SWCNT) and multiple-walled (MWCNT)] and graphs, with carbon quantum-dots (QD) emerging as novel materials for sensor construction in the field of breast and ovarian cancers immunology and cancer research. Mucin (MUC) family of glycoprotein have been investigated as some biomarkers. Ovarian cancers have the worst prognosis when malignancies are diagnosed at late stage, therefore finding biomarkers that help diagnose this cancer at early-stage is essential for patient's survival. Mucins are large glycoproteins. They have a crucial role in the protection and lubrication of the surface of epithelial tissues including mammary glands, female reproductive tracts, kidneys, lungs, stomach, pancreas, or gall bladder. They also play an important role in cell differentiation, signaling processes and cell adhesion. Normal amount of mucin presents in the body, but in ovarian cancer it's overproduction has been seen in the blood. Therefore, measuring this marker in blood can help in the rapid diagnosis of cancer. Cristea et al. [45] have investigated the development of a diagnosis device based on the optical and electrochemical principles for detecting the levels of multiple MUC-type biomarkers in the serum samples of patients. Combining SPR and electrochemical techniques, the selectivity has improved the sensitivity and reliability of the obtained transducer. They have shown the combination of these methods is more accurate than separate ones and it is more cost-effective. This method became more popular through the use of carbon nanotubes. It has excellent properties such as a large surface area, high conductivity, and easy chemical modification as an electrode surface for use in electrochemical immunosensors. Carbon nanotubes also have high chemical and thermal stability; therefore, it can be efficiently used in sensor surface modification. To increase the sensitivity, Rusling's group has developed nanostructured electrodes including densely packed films of single-wall carbon nanotube (SWCNT) sections (forests) [46]. It is ultrasensitive in detecting PSA. They have also identified that by attaching the tracer antibody to these carbon nanotubes with multiple HRP labels, giving a DL of 4 pg mL^{-1} for PSA spiked into undiluted calf serum can enhance the sensitivity even more. In a 4-electrode SWCNT forest array for detection of multiple protein biomarkers [PSA, prostate-specific membrane antigen (PSMA), IL-6 and platelet factor 4 (pf-4)] this technique has investigated patient's serum samples. The results were positive and they have detected its accuracy by ELISA. SWCNT forests can capture immobilized antibodies 10–15-fold more than usual immunosensors, therefore, providing significant enhancements to the sensitivity of the immunosensor. These results indicate that multiply-labeled detection probes that have been used in conjunction with electrode surface modified increase their ability to capture a high-density antibody in immunosensors. A comparison of immunosensors under similar assay conditions to Rustling's group in detection of IL-6 has revealed that the AuNP modified platform yielded a three-fold better detection limit than a SWCNT-based method [46].

Although this technique is acceptable, its accuracy does not exist for all fluids, and other techniques in this area need to be combined with a method like magnetic bead-based immunoassays. The use of magnetic beads to develop diagnostic devices has become more popular. For example, magnetic beads have been used as substrates for the capture of antibodies or for target antigens in immunoassays and enzyme-linked immunoassays. Magnetic beads have several positive features, making it a popular method. These include fast reaction kinetics compared to bulk-solid surfaces, high surface area per unit volume (owing to their small diameter), and good stability. Moreover, the relative ease of surface modification with functional groups, DNA, enzymes, or antibodies greatly contributes to the utility of beads in the development of sensitive, rapid electrochemical immunoassay systems [41–44].

In the latest advancements in research, many immunosensors have been investigated. And common among all of these are their high sensitivity in biomarker detection. The detection of multiple biomarkers in a noninvasive manner will be key in facilitating prompt early-stage cancer diagnosis and reducing associated morbidity and mortality.

3.5   Combination of imaging and immunodiagnostics

It is known that unusual genomic alteration in the levels of nucleic acids and chromatin cause to overexpression of proteins that has responsible to cell function. These proteins might make a patterns that are relevant to malignant progression. In recent studies, it has been seen that greater changes in nucleic acid expression has occurred in end-stage patients with cancer. Therefore, identifying of changes in nucleic acids are essential for cancer screening and monitoring. In addition, it has been reported that some biomarkers such as EGFR, P53, cell regulatory protein Bcl-29, and cell division cycle protein CDK-110 are overexpressed in specific cancers, and they may be significant factors for specific cancers (such as lung cancer, breast cancer, and ovarian cancer). As their roles in cancer have been identified, the detection and evaluation of these nuclear biomarkers (nucleic acids and proteins) will help to clarify the signaling pathways in tumorigenesis, and thus improve early-stage cancer diagnosis [47].

Imaging is some ways to diagnose or to investigate the progression or improvement of the disease. In cancer diagnosis, the goal of imaging is to detect malignant cells from normal cells. Six techniques for assessing diseases are in the imaging subdivisions, including CT, MRI, PET, US, SPECT, and visual detectors. In cancer diagnosis only four of these techniques due to three-dimensional imaging feature which they have can help in detection of cancer cells. As it has been mentioned, imaging is designed for various illnesses, and it's not designed for cancer cell detection alone. This technique doesn't have ability to detect small tumor cells, so the diagnosis of cancer requires designing and modifying these techniques or combining them with new techniques for rapid, sensitivity and high selectivity detection [47,48].

As the specificity and accuracy of cancer investigation is critical, recent studies focus more on the combined use of imaging and other immunodiagnostic assays for maximum effectiveness.

Molecular imaging is a novel technique that has the potential to enhance the diagnostic and therapeutic approaches for cancer treatment. Historically for solid tumors, the methods were used regular imaging modalities such as ultrasonography (US), CT, MRI, and PET. These were not accurate in detecting the biochemical or molecular stage of the neoplastic cell. Molecular Imaging also enhanced the inducible "smart" molecular MRI probes because of their potential for imaging tumors before and after surgical removal, without a renewal of the probe to evaluate the accuracy of surgical resection. Moreover, these smart MR/optical imaging probes can be used in screening programs for the detection of small tumors. New methods have been investigated today and this includes the dual labeling of molecular MR probes with fluorescent dyes. By the development of this technique detection of cancerous tissue from healthy tissue become more facile.

In general, the main aim of scientists is detecting malignancies at an earlier state. In imaging methods, this aim has led to development in techniques such as US, CT, and MRI, however they could not delineate the detection of solid tumors [49]. Molecular imaging techniques vastly improve cancer diagnosis through displaying entirely new possibilities for early detection and effective treatment of cancer; two factors that are crucial to successfully fighting the disease. Molecular imaging is commonly defined as a noninvasive method of imaging cellular and subcellular occurrences. The operation of molecular imaging is based on the distinctive molecular characteristics of malignant cells. Molecular imaging probes detect and highlight these specific characteristics which can be exhibited in, or on any side of each malignant cells, or in the rounding of extracellular matrix and cells in the vicinity, such as T cells, dendritic cells, fibroblasts, macrophages, or endothelial cell. Therefore, molecular imaging of these malignant cells enhances the cancer diagnosis and staging, and makes it easier for tumor detection. Traditional imaging technique such as US, CT, and MRI do not have this ability, however today, detection is continuously developed through the use of molecular imaging.

With all of these new techniques, tumor detection is still based on anatomical characteristics. Recent studies have shown that molecular imaging can detect carcinogenesis at a much earlier time as it can detect alterations on the cellular level, and they are targeted as soon as they occur. For example, signaling changes in malignant cells' glucose metabolism occur early in carcinogenesis and it is detectable by molecular imaging. PET imaging with F-FDG is another technique for detection of these changes, however compared to molecular imaging, it has a limited ability to detect them. For advancements in this technique, further development of more specific molecular imaging probes that target pathologic characteristics, ideally for imaging modalities without radiation exposure for the patient, is necessary. Tumor detection without the need of invasive procedures like biopsies or surgery is considered as another feature of molecular imaging. However, this detection is very difficult for traditional imaging techniques such as CT and MRI. It has been shown that Imaging with PET technique doesn't have the ability to detect tumor cells at its early stage while molecular imaging due to high sensitivity can detect these small tumor cells therefore, ultimately, molecular imaging may be able to determine the best treatment.

Early detection, accurate staging, and complete surgical removal are crucial in order to successfully treat and potentially cure patients with solid cancers. The combination of imaging and immunodiagnostics can provide the necessary techniques to improve cancer diagnoses and could allow for cost-effective personal treatment approaches for patients. Molecular imaging techniques, PET and other techniques are continuously advancing to create effective noninvasive techniques for cancer diagnosis [47–49].

3.5.1   Urine as a noninvasive body fluid

Noninvasive methods in all areas of medicine are important. Stem cell research remains an interesting and exciting area that possesses the potential of improving healthcare for human beings [50–52]. Although stem cells are derived from bone marrow or adipose tissue and is a known method for IPS, it is an invasive method because it can damage the patient's body. Therefore, IPS isolating cells from urine has been under development, as urine is a good source of stem cells. Several studies show that urine is a good body fluid for cancer detection too. In the recent study in 2018, Woo et al. have investigated the detection of androgen-receptor splice variant 7 (AR-V7) which is associated with castration-resistant prostate cancer, in the RNA of urine sample of patients [53]. They found that it is detectable in urine samples and they also have discovered that AR-V7 transcript levels and the AR-V7/AR-FL ratio in urinary EVs were higher in patients with advanced prostate cancer. This study can be a first report that urine-derived RNA is a reliable source for AR-V7 expression analysis. Therefore, this result can open windows for researchers into liquid biopsies as a noninvasive approach. It can be said that if this kind of receptor is detectable, it must have some protein in which the measurement of its levels can be reliable, and a method for cancer diagnosis or investigation of cancer stage during cancer therapy can be used [54].

In general, cancer is a disease that can be easily treated if diagnosed quickly. Therefore, the use of various techniques for rapid and accurate diagnoses has always been significant. The study of PoAbs and MoAbs, along with immunosensor techniques can accelerate the diagnosis of this disease. Today's knowledge of genomic assays has been incorporated and combined with proteomic assay methods. In addition to these methods, the use of protein detection techniques and the combination of these methods have helped imaging techniques such as MRI and CT.

In cancer diagnostic methods, aside from the cost, accuracy, and trustworthiness, other factors are important too, in particular the noninvasiveness of the method. The usage of body fluids such as oral saliva, urine, and blood, is more attractive for detection of biomarkers with high sensitivity and specificity, and they are also far less invasive for patients. Choose of urine as a body fluid with an immunoassay technique that could detect more accurate biomarkers for specific cancer could be a great advance in cancer diagnosis at its early stage.

After introducing the role of immune assays in cancer diagnosis in this chapter, the landmark effect of immunotherapy in cancer will be discussed in the following.

References

[1] R. Etzioni, N. Urban, S. Ramsey, M. McIntosh, S. Schwartz, B. Reid, et al. Early detection: the case for early detection, Nat. Rev. Cancer 3 (4) (2003) 243.

[2] C. Desmetz, A. Mange, B. Maudelonde, J. Solassol, Autoantibody signatures: progress and perspectives for early cancer detection, J. Transl. Med. 15 (10) (2011) 2013–2024.

[3] S. Sharma, R. Raghav, R. O'Kennedy, S. Srivastava, Advances in ovarian cancer diagnosis: a journey from immunoassays to immunosensors, Enzyme Microb. Technol. 89 (2016) 15–30.

[4] H. Prasetya, B.B. Purnomo, I.K.G. Muliartha, S.R. Prawiro, Polyclonal antibody from 47 kDa protein of bladder cancer is sensitive and specific for detection of bladder cancer, Biomark. Genom. Med. 6 (3) (2014) 116–120.

[5] H. Prasetya, B.B. Purnomo, K. Mintaroem, S.R. Prawiro, Sensitivity and specificity of 47kDa polyclonal antibody for detection of bladder cancer cells in urine of hematuria patients, Afr. J. Urol. 24 (4) (2018) 264–269.

[6] A. Singh, S. Chaudhary, A. Agarwal, A.S. Verma, Antibodies: monoclonal and polyclonal, Animal Biotechnology, Academic Press, (2014) pp. 265–287.

[7] D.J. Brennan, D.P. O'connor, E. Rexhepaj, F. Ponten, W.M. Gallagher, Antibody-based proteomics: fast-tracking molecular diagnostics in oncology, Nat. Rev. Cancer 10 (9) (2010) 605.

[8] S.H. Tan, A. Rastogi, S. Banerjee, A. Bagga, C. Xavier, A. Mohamed, et al. Immunobiomarkers: structural and functional characterization of single chain fragment variable (scFv) to ERG from a mouse monoclonal antibody, 78(13) (2018).

[9] A.M. Wu, Engineered antibodies for molecular imaging of cancer, Methods 65 (1) (2014) 139–147.

[10] G. Lee, E. Laflamme, C.H. Chien, H.H. Ting, Molecular identity of a pan cancer marker, CA215, Cancer Biol. Ther. 7 (12) (2008) 2007–2014.

[11] G. Lee, Cancer cell-expressed immunoglobulins: CA215 as a pan cancer marker and its diagnostic applications, Cancer Biomark 5 (3) (2009) 137–142.

[12] H. Torikai, A. Reik, P.Q. Liu, Y. Zhou, L. Zhang, S. Maiti, et al. A foundation for "universal" T-cell based immunotherapy: T-cells engineered to express a CD19-specific chimeric-antigen-receptor and eliminate expression of endogenous TCR, Blood 119 (24) (2012) 5697–5705.

[13] G. Lee, M. Zhu, B. Ge, S. Potzold, Widespread expressions of immunoglobulin superfamily proteins in cancer cells, Cancer Immunol. Immunother. 61 (1) (2012) 89–99.

[14] J. Wu, X. Li, W. Song, Y. Fang, L. Yu, S. Liu, et al. The roles and applications of autoantibodies in progression, diagnosis, treatment and prognosis of human malignant tumours, Autoimmun. Rev. 16 (12) (2017) 1270–1281.

[15] C. Desmetz, A. Mange, T. Maudelonde, J. Solassol, Autoantibody signatures: progress and perspectives for early cancer detection, J. Cell. Mol. Med. 15 (10) (2011) 2013–2024.

[16] L. Dai, J.-C.J. Tsay, J. Li, T.-A. Yie, J.S. Munger, H. Pass, et al. Autoantibodies against tumor-associated antigens in the early detection of lung cancer, Lung Cancer 99 (2016) 172–179.

[17] M. Kunizaki, T. Sawai, H. Takeshita, T. Tominaga, S. Hidaka, K. To, et al. Clinical value of serum p53 antibody in the diagnosis and prognosis of colorectal cancer, Anticancer Res. 36 (8) (2016) 4171–4175.

[18] Y. Liu, Y. Liao, L. Xiang, K. Jiang, S. Li, M. Huangfu, S. Sun, A panel of autoantibodies as potential early diagnostic serum biomarkers in patients with breast cancer, Int. J. Clin. Oncol. 22 (2) (2017) 291–296.

[19] L. Wu, W. Chang, J. Zhao, Y. Yu, X. Tan, T. Su, et al. Development of autoantibody signatures as novel diagnostic biomarkers of non–small cell lung cancer, Clin Cancer Res. 16 (14) (2010) 3760–3768.

[20] H.T. Tan, J. Low, S.G. Lim, M.C. Chung, Serum autoantibodies as biomarkers for early cancer detection, FEBS J. 276 (23) (2009) 6880–6904.

[21] Q. Xia, X.T. Kong, G.A. Zhang, X.J. Hou, H. Qiang, R.Q. Zhong, Proteomics-based identification of DEAD-box protein 48 as a novel autoantigen, a prospective serum marker for pancreatic cancer, Biochem. Biophy. Res. Commun. 330 (2) (2005) 526–532.

[22] C.A. Borrebaeck, Precision diagnostics: moving towards protein biomarker signatures of clinical utility in cancer, Nat. Rev. Cancer 17 (3) (2017) 199.

[23] F.X. Sutandy, J. Qian, C.S. Chen, H. Zhu, Overview of protein microarrays, Curr. Protocols Protein Sci. 72 (1) (2013) 27.1.

[24] H.D. Shukla, Comprehensive analysis of cancer-proteogenome to identify biomarkers for the early diagnosis and prognosis of cancer, Proteomes 5 (4) (2017) 28.

[25] B.S. Shruthi, P. Vinodhkumar, Selvamani, Proteomics: a new perspective for cancer, Adv. Biomed. Res. 5 (2016) 67.

[26] P. Mertins, D.R. Mani, K.V. Ruggles, M.A. Gillette, K.R. Clauser, P. Wang, et al. Proteogenomics connects somatic mutations to signalling in breast cancer, Nature 534 (7605) (2016) 55.

[27] K.S. Anderson, N. Ramachandran, J. Wong, J.V. Raphael, E. Hainsworth, G. Demirkan, et al. Application of protein microarrays for multiplexed detection of antibodies to tumor antigens in breast cancer, J. Proteome Res. 7 (4) (2008) 1490–1499.

[28] C. Wingren, A. Sandström, R. Segersvärd, A. Carlsson, R. Andersson, M. Löhr, C.A. Borrebaeck, Identification of serum biomarker signatures associated with pancreatic cancer, Cancer Res. 72 (10) (2012) 2481–2490.

[29] H. Srinivasan, Y. Allory, M. Sill, D. Vordos, M.S.S. Alhamdani, F. Radvanyi, et al. Prediction of recurrence of non muscle-invasive bladder cancer by means of a protein signature identified by antibody microarray analyses, Proteomics 14 (11) (2014) 1333–1342.

[30] M. Puig-Costa, A. Codina-Cazador, E. Cortés-Pastoret, C. Oliveras-Ferraros, S. Cufí, S. Flaquer, et al. Discovery and validation of an INflammatory PROtein-driven GAstric cancer Signature (INPROGAS) using antibody microarray-based oncoproteomics, Oncotarget 5 (7) (2014) 1942.

[31] X. Quan, Y. Ding, R. Feng, X. Zhu, Q. Zhang, Expression profile of cytokines in gastric cancer patients using proteomic antibody microarray, Oncol. Lett. 14 (6) (2017) 7360–7366.

[32] J.M. Schwenk, U. Igel, M. Neiman, H. Langen, C. Becker, A. Bjartell, F. Ponten, F. Wiklund, H. Gronberg, P. Nilsson, et al. Toward next generation plasma profiling via heat-induced epitope retrieval and array-based assays, Mol. Cell. Proteomics 9 (11) (2010) 2497–2507.

[33] J.C. Miller, H. Zhou, J. Kwekel, R. Cavallo, J. Burke, E.B. Butler, et al. Antibody microarray profiling of human prostate cancer sera: antibody screening and identification of potential biomarkers, Proteomics 3 (1) (2003) 56–63.

[34] S. Darmanis, T. Cui, K. Drobin, S.C. Li, K. Oberg, P. Nilsson, J.M. Schwenk, V. Giandomenico, Identification of candidate serum proteins for classifying well-differentiated small intestinal neuroendocrine tumors, PLoS One 8 (11) (2013) e81712.

[35] S.A. Khan, D. Reddy, S. Gupta, Global histone post-translational modifications and cancer: biomarkers for diagnosis, prognosis and treatment?, World J. Biol. Chem. 6 (4) (2015) 333–345.

[36] Z. Chen, T. Dodig-Crnković, J.M. Schwenk, S.C. Tao, Current applications of antibody microarrays, Clin. Proteomics 15 (2018) 7, doi: 10.1186/s12014-018-9184-2.

[37] P. Delfani, L. Dexlin Mellby, M. Nordström, A. Holmér, M. Ohlsson, C.A.K. Borrebaeck, et al. Technical advances of the recombinant antibody microarray technology platform for clinical immunoproteomics, PLoS One 11 (7) (2016) e0159138.

[38] B. Bohunicky, S.A. Mousa, Biosensors: the new wave in cancer diagnosis, Nanotechnol. Sci. Appl. 4 (2010) 1–10.

[39] F.S. Felix, L. Angnes, Electrochemical immunosensors—a powerful tool for analytical applications, Biosens. Bioelectron. 102 (2018) 470–478.

[40] V.S.A. Jayanthi, A.B. Das, U. Saxena, Recent advances in biosensor development for the detection of cancer biomarkers, Biosens. Bioelectron. 91 (2017) 15–23.

[41] Su, Li, et al. Development of novel piezoelectric biosensor using pzt ceramic resonator for detection of cancer markers. *Biosensors and biodetection.* Humana Press, New York, NY, 2017, pp. 277–291.

[42] S. Sharma, H. Byrne, R.J. O'Kennedy, Antibodies and antibody-derived analytical biosensors, Essays Biochem. 60 (1) (2016) 9–18.

[43] E. Spain, S. Gilgunn, S. Sharma, K. Adamson, E. Carthy, R. O'Kennedy, R.J. Forster, Detection of prostate specific antigen based on electrocatalytic platinum nanoparticles conjugated to a recombinant scFv antibody, Biosens. Bioelectron. 77 (2016) 759–766.

[44] E. Arkan, R. Saber, Z. Karimi, M. Shamsipur, A novel antibody–antigen based impedimetric immunosensor for low level detection of HER2 in serum samples of breast cancer patients via modification of a gold nanoparticles decorated multiwall carbon nanotubeionic liquid electrode, Anal. Chim. Acta 874 (2015) 66–74.

[45] C. Cristea, A. Florea, R. Galatus, E. Bodoki, R. Sandulescu, D. Moga, D. Petreus, Innovative immunosensors for early stage cancer diagnosis and therapy monitoring, in: The International Conference on Health Informatics, Springer, Cham, 2014, pp. 47–50.

[46] J.F. Rusling, G.W. Bishop, N. Doan, F. Papadimitrakopoulos, Nanomaterials and biomaterials in electrochemical arrays for protein detection, J. Mater. Chem. B 2 (1) (2014)doi: 10.1039/C3TB21323D.

[47] J.P. O'connor, E.O. Aboagye, J.E. Adams, H.J. Aerts, S.F. Barrington, A.J. Beer, D.L. Buckley, Imaging biomarker roadmap for cancer studies, Nat. Rev. Clin. Oncol. 14 (3) (2017) 169.

[48] T. Hussain, Q.T. Nguyen, Molecular imaging for cancer diagnosis and surgery, Adv. Drug Deliv. Rev. 66 (2013) 90–100.

[49] Y.C. Chen, X. Tan, Q. Sun, Q. Chen, W. Wang, X. Fan, Laser-emission imaging of nuclear biomarkers for high-contrast cancer screening and immunodiagnosis, Nat. Biomed. Eng. 1 (9) (2017) 724.

[50] N. Pavathuparambil Abdul Manaph, M. Al-Hawwas, L. Bobrovskaya, P.T. Coates, X.F. Zhou, Urine-derived cells for human cell therapy, Stem Cell Res. Ther. 9 (1) (2018) 189.

[51] A. Di Meo, J. Bartlett, Y. Cheng, M.D. Pasic, G.M. Yousef, Liquid biopsy: a step forward towards precision medicine in urologic malignancies, Mol. Cancer 16 (1) (2017) 80.

[52] X. Ji, M. Wang, F. Chen, J. Zhou, Urine-derived stem cells: the present and the future, Stem Cells Int. 2017 (2017) 4378947.

[53] H.K. Woo, 1, J. Park, J.Y. Ku, C.H. Lee, V. Sunkara, H.K. Ha, Y.K. Cho, Urine-based liquid biopsy: non-invasive and sensitive AR-V7 detection in urinary EVs from patients with prostate cancer, Lab Chip 19 (1) (2018) 87–97.

[54] G. Siravegna, S. Marsoni, S. Siena, A. Bardelli, Integrating liquid biopsies into the management of cancer, Nat. Rev. Clin. Oncol. 14 (9) (2017) 531.

Immunotherapy

4

Chapter outline

Bio-Engineering Approaches to Cancer Diagnosis and Treatment. http://dx.doi.org/10.1016/B978-0-12-817809-6.00004-2

4.1   Cancer immunotherapy

Cancer is the principal cause of early death worldwide. American Cancer Society has appraised that nearly 1.7 million new cancer cases and 609,640 deaths happened in 2018 in the US. The mortality rates due to cancer types include lung (1.69 million), liver (788,000), colorectal (774,000), stomach (754,000), and breast (571,000) cancers [1]. Common cancer remedies such as tumor surgery, chemotherapy, and hormonal treatments are subject to limitations. For instance, tumor surgery cannot inhibit metastasis, and radiation therapy is costly and time consuming. The anticancer drugs in chemotherapy can be rapidly released throughout the body and are toxic to noncancerous cells. Moreover this approach is ineffective in eliminating metastatic cancer cells [2].

Thomas and Burnet have demonstrated that the immune system can contribute to the control and progression of cancer [3]. Cancer immunotherapy has evolved to promote a cancer-specific immune response to eliminate tumor cells as a new treatment method. It has been shown that the immune system can suppress tumors by apoptosis and phagocytosis of cancer cells.

Immunotherapy can be used in the following cases:

- Slowing and inhibiting tumor cell growth
- Prevent cancer from permeating to other organs of the body
- Improving the immune system function in destroying tumor cells

Different types of cancer immunotherapy include blockade of immune checkpoints, oncolytic virus therapy, T cell therapy, cancer vaccines, antibody based targeted therapy, and cytokine therapy.

4.1.1   Blockade of immune checkpoints

Blockade of immune checkpoints is the most promising approach in activating remedial antitumor immunity. Immune checkpoints are described to a plethora of inhibitory pathways that are essential for retaining self-tolerance and regulating the duration and amplitude of physiological immune responses in order to reduce tissue damage. It is now explicit that tumors adjust to certain immune-checkpoint pathways as the main mechanism of immune resistance, especially against T cells that are characteristic for tumor antigens. Since many of the immune checkpoints are activated by ligand-receptor interactions, they can be easily blocked by antibodies or regulated by recombination of ligands or receptors. The different immune checkpoints can be referred to as CTLA4, PD-1, PD-L1, LAG 3, B7-H3, B7-H4, CD27, and CD70. Cytotoxic T lymphocyte antigen 4 (CTLA-4), and programmed cell death1/programmed cell death ligand (PD-1/PD-L1) are the two most studied immune check points.

4.1.1.1   CTLA4

The first FDA approved strategy in the immunotherapeutics category was CTLA-4 antibodies [4]. CTLA-4 (also named CD152) is expressed on regulatory T cells (T_{reg}) and modulated on activated CD8+ effector T cells. It works by binding to its ligands, CD80 (B7-1) or CD86 (B7-2) on the surface of antigen presenting cells (APC). The route of T cell activation includes multiple activating signals. CD28, which is expressed on T cells, binds to the same ligands of CTLA-4, providing a principal co-stimulatory signal for subsequent T cell activation after the T cell receptor (TCR) signaling. Moreover, CTLA-4 also participates in the immune regulation of CD4+ T cells. CTLA-4 is expressed on activated CD8+ effector T cells, and seems to operate through two main subtypes of CD4+ T cells: downregulating the activity of helper T cells, and boosting immunosuppression of T_{reg}. Recent research has also demonstrated that the T_{reg}-specific CTLA4 blockade could noticeably reduce the regulation of both auto-immunity and antitumor immunity. Hence, the mechanism of CTLA4 blockade, both the increase of CTLA-4/B7 binding and inhibition of T_{reg}-dependent immunosuppression can be a cancer treatment strategy.

4.1.1.2   PD-1

PD-1, also known as CD279, is one of the main immune checkpoints that restrains an excessive immune response and thus inhibits autoimmunity. The difference of PD-1 with other members of CD28/CTLA-4 family is the absence of an extracellular cysteine residue preventing the creation of covalent dimers [5].

Unlike the CTLA4, the main role of PD-1 is to restrict the activity of T cells in peripheral tissues at the time of an inflammatory response to an infection and to restrict autoimmunity. PD-1 expression is stimulated when T cells become activated. When actuated by one of its ligands, PD-1 impedes kinases that are required in T cell activation through the phosphatase SHP 250, although extra signaling pathways are also most likely stimulated. Also, because PD-1 engagement impedes the TCR 'stop signal,' this pathway could improve the duration of T cell—APC or T cell—target cell contact. PD-1, like CTLA4, is greatly expressed on T_{reg} cells, where their proliferation may be increased in the presence of the ligand. Since many tumors are extremely infiltrated with T_{reg} cells, probably suppress more effector immune responses, blockade of the PD-1 pathway may also increase antitumor immune responses by reducing the number and/or the suppressive activity of intratumoral T_{reg} cells.

The ligands of PD-1 consist of PD-L1 (also known as B7-H1, CD274) and PD-L2 (also known as B7-DC, CD273). However, separate from PD-1, PD-L1 can also connect to B7-1 (CD80) and PD-L2 can connect to repulsive guidance molecule B (RGMb). The expression of PD-L1 on T cells, B cells, endothelial and epithelial cells can be modulated by cytokines such as IFN-γ and TNF-α, which helps retain the peripheral tolerance. The detailed mechanism of the inhibition of PD-L1 and PD-L2 in T cell activity is yet to be clarified. It may involve the enhanced programmed cell death caused by the binding of PD-1 to PD-L1.

4.1.2 Oncolytic viral therapy

Oncolytic viral therapy (OVT) is a new approach for cancer treatment that has potential. Oncolytic viruses (OVs) can duplicate in tumor cells but not in healthy cells, by lysis of the tumor mass. Some viruses have the tendency to infect and kill tumor cells, known as OV, and this category includes viruses available in nature as well as manipulated viruses in the lab to efficiently replicate in tumor cells without damaging normal cells. The use of OVT has long been viewed as an appropriate approach for directly killing cancer cells. The aim of OVT is not only to kill cancer cells but to also activate the immune system that is suppressed by the tumor microenvironment. The biological mechanisms used by viruses to eradicate tumors depend on different factors such as the virus, the target tissue or cell, and the biological pathways targeted. OVT are capable of making long-term memory. Transformed cells can evade from the immune system by transforming their antigens and becoming undetectable to leukocytes in a process named "immunoediting." When OVs infect cancer cells, an inflammatory response is induced. These viruses are able to stimulate immunogenic cell death (ICD) and this form of apoptosis can stimulate an efficient antitumor by activation of dendritic cells (DCs) and the stimulation of specific T lymphocytes.

The first FDA approved OVT was a treatment for a melanoma known as talimogene laherparepvec. This is a modified herpes simplex virus (HSV) that enhances the number of tumor-specific CD8+ T cells and decreases the number of regulatory and suppressor T cells.

Ultimately, there are obstacles that come with the use of this approach. The main obstacle is the host immune response to a viral infection (especially in IV administration) after IV injection, the virus is phagocytized and cannot reach the tumor site. To overcome this problem, accurate design of the optimal viral gene edition and attachment of combinations is needed [6].

4.1.3 Adoptive cell therapy

For the treatment of less immunogenic tumors, such as pancreatic cancer and MSS (microsatellite stable) colorectal cancer, the use of immune checkpoint inhibitors (ICIs) alone is not enough. Consequently, collecting a population of tumor-reactive T cells through adoptive cell therapy (ACT) would make immune-based therapies possible [7]. ACT is a cancer therapy strategy that uses either natural host cells that show antitumor reactivity, or host cells that have been genetically engineered with antitumor TCRs or chimeric antigen receptors (CARs). ACT has numerous benefits compared to other types of cancer immunotherapies, depends on the active in vivo development of adequate numbers of antitumor T cells with the functions essential to mediate cancer eradication. Perhaps the most important point, using this approach is allowing the manipulation of the host before a cell transfer in order to create a more suitable microenvironment that is advantageous in supporting antitumor immunity.

ACT does not only consist of engineered CAR-T cells, but also other types of ACT including endogenous T cell (ETC) and tumor-infiltrating lymphocyte (TIL) therapy. With the use of these methods, ACT has been utilized to treat a variety of cancers including melanoma, cervical cancer, lymphoma, leukemia, bile duct cancer, and neuroblastoma [8].

4.1.3.1 CAR

The main restriction of the effective use of ACT is the recognition of cells that can target antigens selectively expressed on the tumor and not healthy cells.

In the lab, a specific receptor which was planned to bind with predetermined proteins on tumor cells was produced. The CAR is then attached to T cells. This aids the T cells to detect and kill tumor cells that have the defined protein that the receptor is set to bind with. These modified T cells (CAR T cells) are then proliferated in large numbers in the lab and conveyed to patients with cancer. The FDA-approved CAR T cells are made of an extracellular single chain Fv (scFv) antibody fragment led against the B cell-specific antigen CD19, the transmembrane domain and intracellular TCR activation and co-stimulatory domains [9].

4.1.3.2 ETC

ETC therapy developed as a treatment strategy where endogenous tumor-reactive T cells can be found in the peripheral blood of patients, isolated and proliferated while keeping antitumor activity. Some groups have been recapitulated in vivo biology of T cell induction, using autologous APCs (e.g., DCs) or artificial APCs that can convey an antigen-specific TCR signal along with essential costimulatory ligands to enhance the number of low frequency antigen-specific T cells in the blood. The use of TCR in the peripheral blood creates a broader net in obtaining scarce antigen-specific T cells, many of which are targeted to a shared tumor-associated self-antigen, and unlike TIL therapy, does not necessitate TIL-rich samples which may not be available or may be difficult to surgically access.

Another biological benefit to the utilization of peripheral blood as a source of effector T cells is:

- The capacity to produce longer lasting T cells from a nonmanipulated population
- Making a rich source of extremely replicative, 'helper-independent' central memory type CTL

The use of this approach has exhibited clinical efficiency in the treatment of patients with refractory metastatic melanoma and other solid tumor malignancies, albeit in relatively few trials.

4.1.4 Tumor infiltrating lymphocyte

TIL therapy utilizes the patient's T cells which are gathered from a section of surgically removed tumors. Although these cells may detect the cancer, few of them are usable. The number of these cells is proliferated and treated (with the aim of active the lymphocytes) in the lab and then returned to the patient's body to help their immune system eradicate cancer cells.

Melanoma, among tumors with infiltrating lymphocytes, is one of the first examples where tumor reactive T cells could be increased in vitro to a large population for adoptive transfer. For nonmelanoma tumors, the use of novel methods such as in vitro activation of costimulatory ligands, or selection of PD1+ tumor reactive TIL has shown efficient preclinical results for ovarian, breast, and colorectal cancers.

In addition to the benefits, this method also handles limitations such as toxicity due to intense preinfusion and postinfusion regimens, and inappropriate patient selection that limits the use of this approach [7].

4.2 Antibody based targeted therapy

Antibody based targeted therapy as a method for treating cancer has been invented since 40 years ago and is now one of the most important and successful ways to treat patients with hematological malignancies and solid tumors. The basis of this method refers to observation of expression of antigens by tumor cells using serological techniques in 1960 [10]. Subsequently, in 1975, the production of monoclonal antibodies was developed using the hybridoma technology by Milstein and Köhler [11]. Identification and production and use of antibodies was continued until 1997 when rituximab was introduced as the first full length chimeric antibody and used to treat Patients with Relapsed low-grade non-Hodgkin's lymphoma [12]. Following this, scientists were drawn to the use of antibodies and antibody based targeted therapy as a solution to treat malignancy. So today, antibody based targeted therapy, that is, the use of whole antibody or specific antibody fragments; alone or in combination with drugs, toxins, radionuclides, etc. is very common and used in clinical practice.

In order to better understand the function of antibodies to treat malignancy, there is a need to briefly define malignancy. There are several basic question in this regard: What is cancer? What is the difference between benign tumor and malignant tumor? What is the meaning of metastasis? In simple terms, normal cells grow and divide slowly and there are some components in the cell's structure that control them. Cancer occurs when the cells change, quickly grow and divide without any control. When the tumor is formed, if it continues to grow unbroken and invades the surrounding tissue, called a malignant tumor; a malignant tumor can enter the blood vessels or lymph nodes and go to other parts of the patient's body, where it begins to grow and divide and create a new tumor. This cancer spread in the body is called metastasis.

One of the most important goals for cancer treatments is to prevent malignant activity of the cancer cells without affecting the activity of healthy cells. And because one of the important features of the method of antibody based targeted therapy is selectivity, in other words, the ability to detect and remove tumor cells and destroy them without damaging normal cells; this method has been highly respected by researchers.

4.2.1   The mechanism of antibody based targeted therapy

Antibodies exhibit therapeutic effects by targeting the surface expressed antigens on cancer cells (Fig. 4.1). Before investigating the mechanisms of antibody activity on cancer cells, it is necessary to briefly describe their characteristics, while we were covered them in detail in the previous sections.

Antibodies, commonly used in form of immunoglobulins, are proteins that have two binding arms. Antigen-binding portion and *Fc-region* that is responsible for binding to serum proteins (complement) or cells. The use of antibodies for treatment of cancer is divided into two general methods:

-   An antibody directly and by alterations in intracellular signaling or inhibition of function of growth factor receptors or inhibition of function of adhesion molecules causes tumor cell apoptosis [13].
-   Antibody itself does not have the responsibility to kill the target cells, but instead marks the target cells that other components or effector cells of the body's immune system can attack the cell. The mechanisms for attacking the labeled cells with the antibodies are made from two paths:
-   Antibody dependent cellular cytotoxicity (ADCC)
-   Antibody-dependent complement mediated cytotoxicity (CMC)

ADCC includes the identification of the antibodies by immune system cells that interact with antibody marked cells directly or through other cells (Fig. 4.2). And CMC is a process in which different complement proteins act as cascades; it usually happens when there are several immunoglobulins in close proximity to each other that either directly lead to cell lysis or indirectly invoke other immune system cells to this location [14].

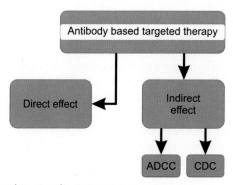

FIGURE 4.1 The mechanism of antibody based targeted therapy.

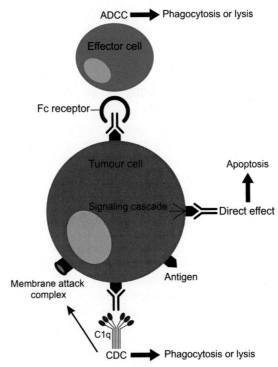

FIGURE 4.2 Identification of the antibodies by immune system cells.

Cell surface differentiation antigens are a different group of glycoproteins and carbohydrates that can be seen at the surface of normal and tumor cells. The antigens that participate in signaling for growth and differentiation are often growth factors and growth factor receptors. For example, CEA2, epidermal growth factor receptor (EGFR; also known as ERBB1), ERBB2 (also known as HER2), ERBB3, MET (also known as HGFR), insulin-like growth factor 1 receptor (IGF1R), ephrin receptor A3 (EPHA3), tumor necrosis factor (TNF)-related apoptosis-inducing ligand receptor 1 (TRAILR1; also known as TNFRSF10A), TRAILR2 (also known as TNFRSF10B) and receptor activator of nuclear factor-κB ligand (RANKL; also known as TNFSF11) are growth factors that targeted by antibodies in the treatment of patients with cancer. And the antigens involved in angiogenesis are usually proteins or growth factors such as: vascular endothelial growth factor (VEGF), VEGF receptor (VEGFR), integrin αVβ and integrin α5β1 [15]. For example, regarding the use of antibodies to target a specific antigen in patient with cancer, Patel et al. have used Cetuximab associated poly(lactide-co-glycolide) nanoparticles to target epidermal growth factor receptor which extremely expressed in nonsmall lung cancer cells [16]. And in another example in 2018, Carolin Offenhäuser et al. in a study conducted in vitro and then in an animal model of mice, have shown that targeting EphA3 using antibody-drug conjugate (ADC) and radioimmunotherapy (RIT) methods can be effective in treating glioblastoma. The antibody they used in this study is IIIA4, which is a EphA3 monoclonal antibody (mAb) [17].

Antibody based targeted therapy may be the most effective way to treat various malignancies and cancers. But the basic point is that most of these antigens are not expressed only in tumor cells, but also expressed in healthy cells! It is important to state that today the problem with therapies for the treatment of cancer is not the destruction of cancer cells, but the destruction of cancer stem cells, and due to the heterogeneity of cancer stem cells, and since these cells express similar markers to noncancer stem cells, it is not possible to detect these two cell types using a specific marker. And this makes it difficult to reach the goal of antibody based targeted therapy (which eliminates tumor cells without affecting healthy cells). As an instance CD105 is known as the proliferation marker for endothelial cells and it is overexpressed on tumor neo vasculature, but also expressed in normal cells.

In this regard, attention should be paid to the specific epitopes of the antigens that can be different in tumor cells and normal cells or a more detailed and specialized examination of which molecules are expressed in cancer cells at the cell surface, while expressed in healthy cells.

4.2.2 Whole antibody or antibody fragments

Antibody fragments are small and simple structure that today is highly regarded because of the many advantages they have over the use of whole antibodies. Antigen-binding fragments (Fab) and single chain variable fragments (scFV) are common antibody fragments that have been investigated and also another type known as "third generation" (3G) molecules. The Fab fragments are consisting of one constant and one inconstant domain of heavy and light chains, whereas in scFV fragments, the varying areas of heavy and light chains are merged and the constant heavy and light chain that was in the previous state is not here and in the case of the third type, it consists of only one variable heavy chain (Fig. 4.3) [18,19].

As already mentioned whole antibodies are much more successful in treating hematological malignancies than solid tumors, and this is due to the size of antibodies (150 kDa) that are restricted to penetrate solid tumors. Lack of influence of antibodies to solid tumors can be solved using smaller antibody fragments (it is about 50 kDa for Fab and 25 kDa for scFV) [20]. Unlike whole antibodies, the antibody fragments do not compete with endogenous immunoglobulins or does not bind to FC receptors on healthy cells due to lack of FC portion and other benefits of the absence of FC portion, is reducing the immunogenicity potential of antibody fragments based drugs. Another advantage would prefer antibody fragments to whole antibodies is more sensitivity in antigen detection [21]. Considering the advantages mentioned above for using antibody fragments instead of whole antibodies, the attention of researchers has been drawn to the use of antibody fragments in recent years. For example, in a study that was carried out in 2016 by Mazzocco et al. anti-PSMA (prostate-specific membrane antigen) scFv fragment is used to diagnose prostate cancer. They have labeled ScFvD2B that is an anti-PSMA

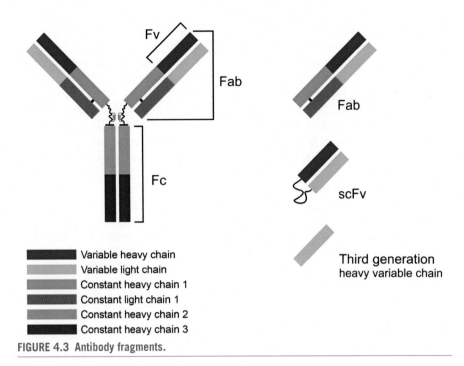

Variable heavy chain
Variable light chain
Constant heavy chain 1
Constant light chain 1
Constant heavy chain 2
Constant heavy chain 3

FIGURE 4.3 Antibody fragments.

antibody fragment, using fluorescence to produce a specific imaging probe. Then this labeled ScFvD2B injected into the mice having prostate tumors and then, after studying fluorescence, they have seen that these labeled anti-PSMA scFv fragment are closely connected to PSMA. They have shown that ScFvD2B can be a suitable agent for detection of PSMA-expressing cells in patients suspected to having prostate cancer [22].

4.2.3 Immunoconjugates and unconjugated antibody

Antibodies are used in both forms of immunoconjugates or unconjugated form to fight against tumor cells. Unconjugated or naked antibody by the methods described in Fig. 1 generally results in the destruction of tumor cells. While in the immunoconjugates antibody, antibodies are associated with radionuclides, drugs or toxins; and carry them to the target. As indicated at the beginning of the chapter, the most important goal of cancer therapies is to destroy tumor cells without harming or at least effect on healthy cells. Immunoconjugates illustrate an important tool on the way to this direction. The components of the immunoconjugates method are: antibody as carrier, a cytotoxic agent (i.e., divided into three categories: radionuclides, drugs, and toxins) and a chemical linker which attaches this cytotoxic agent to antibody [23]. The advantage and disadvantages of each of these three cytotoxic agents are discussed in detail later in this chapter.

4.2.4　Clinical examinations

Norton et al. compared the use of chemotherapy and chemotherapy associated with trastuzumab as a treatment for her2-positive breast cancer patients. By examining the serum of patients before and after treatment, they observed that before treatment, all patients had the same mean anti-HER2 IgG levels; but after treatment, the mean levels of antibodies in the group receiving chemotherapy with trastuzumab has been significantly higher. Whereas in the group that has just been chemotherapy, there has been no change in the level of antibodies [24].

In the other clinical research Chari et al. have evaluated safety and tolerability of daratumumab in combination with pomalidomide/dexamethasone (pom-dex) in patients with multiple myeloma. They have found that compared to other therapies, as with the use of pom-dex alone and without combining whit daratumumab, there is no significant increase in the rate of infection, despite the high rate of neutropenia seen. They also found that the use of daratumumab together with pom-dex induces rapid, perdurable and definite response in patients who received this treatment [25].

4.3　Radio immunotherapy (RIT)

4.3.1　Radiotherapy

Radiotherapy is one of the major therapies for cancer treatment. Radiation therapy or oncology radiation is the use of ionized rays (these rays are ionized by passage of matter and produce positive and negative ions in the material) as an agent for treating cancer and controlling or killing malignant cells. Radiotherapy can treat some of cancers that limited to part of the body. It can also be treated as part of the treatment and prevent tumor recurrence after surgery by removing malignant tumors, for example, prostate carcinomas, cervix carcinomas, lung carcinomas (nonsmall cell) and lymphomas (Hodgkin's and low grade non-Hodgkin's) are considered to be cancers that can be treated with radiotherapy alone. And breast carcinomas, rectal and anal carcinomas, advanced lymphomas and CNS tumors are among the many cancers that are curable with radiotherapy in combination with other treatment such as surgery, chemotherapy, hormone therapy, and immune therapy. The mechanism of radiotherapy as a treatment for cancer is to damage the DNA of tumor cells. The damage can directly or indirectly affect the DNA chain; more clearly, radiation can directly affect DNA inside the cell and cause it to degrade or that radiation can cause ionization of intracellular components and cause free radicals and thus indirectly damages DNA (Fig. 4.4). On the other hand, this damage can be entered into a DNA strand or entered into two strands of DNA. And since the cells have a mechanism for repairing single strand DNA breaks (SSDB), it has been seen that double strand DNA breaks (DSDB) is the most important technique that results in cell death [26].

Radiotherapy as a treatment for cancer induces different types of cell death. Apoptosis and mitotic cell death or mitotic catastrophe are the major cell deaths which happens after radiotherapy. In addition, radiation may cause other types of

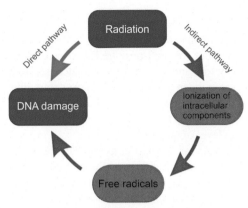

FIGURE 4.4 **The mechanism of radiotherapy as a treatment for cancer and damage the DNA of tumor cells.**

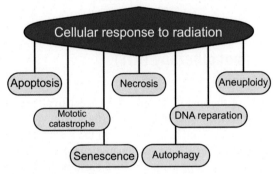

FIGURE 4.5 Cellular response to radiation.

cell death, such as senescence, necrosis and autophagy. On the other hand, the cell may give another response to radiation. In other words, the cell may not die and repair the damage DNA. Another response that the cell may give to radiation is when factors such as p53 are not activated and aneuploidy occurs and the tumor progresses [26,27]. Cellular response to radiation; apoptosis and mitotic catastrophe are the main cellular responses to radiation while the cell may give other responses to the radiation as shown in Fig. 4.5.

Despite the fact that radiotherapy has always been considered as one of the effective strategies for cancer treatment, several questions are raised for researchers: what factors affect cellular response to radiation? And how can we control cellular response to radiotherapy? Or how can radiation be limited to cancer cells? And how can prevented the effect of radiation on normal tissues? Or can combination of radiotherapy with other therapies be more effective? And some other question like these, that have led to better and more effective treatments for cancer treatment. One of these effective methods is radio immune therapy (RIT), which is discussed in the next section.

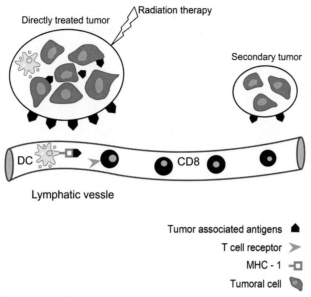

FIGURE 4.6 The abscopal effect.

4.3.1.1 Radiotherapy and its effect on the immune system

As noted in the previous section, radiotherapy destroys the tumor locally, and this is done by destruction the DNA. Along with this local action that destroys the DNA of tumor cells, radiotherapy can cause changes in the immune response against cancer antigens [28]. Significantly, radiotherapy works not only against a localized irradiated tumor, but also against nonirradiated tumor that is created by metastasis at a distance from the primary tumor. This phenomenon is called abscopal and was introduced for the first time by Mole in 1953 [29]. The abscopal effect is schematically shown in Fig. 4.6.

As shown in Fig. 4.6, in the interpretation of the abscopal effect, the radiation generates tumor associated antigens (TAAs). These TAAs are then identified by APCs such as DCs and connected to them. This connection is made by the major histocompatibility complex class 1 (MHC-1), and DCs introduce TAAs to T cells in this way. Another consequence of radiotherapy is the activation and maturation of DCs which occurs when damage-associated molecular patterns (DAMPs) such as uric acid, ATP, high-mobility group box 1 (HMGB1), heat shock protein, IL-1α and hyaluronic acid are released and detected by Toll-like receptors (TLRs) at the surface of DCs [30]. In addition to the beneficial effects that radiotherapy has on stimulating the immune system to fight cancer cells, it also has inhibitory effects on immune system. For example, radiotherapy can increase regulatory T-cells (T_{reg} cells) in the tumor microenvironment and accumulation of T_{reg} cells in that microenvironment and secretion of some cytokines such as IL-10 and TGFβ can inhibit myeloid-derived suppressor cells (MDSCs) [31]. Therefore, in contrast to the beneficial stimulatory

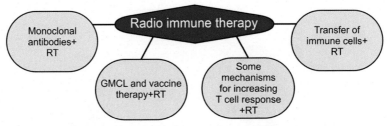

FIGURE 4.7 Four main categories of radio immune therapy (RIT).

effects that radiotherapy has on the immune system, there are suppressive and inhibitory effect, which in general can be said to be somewhat equivalent. As a result, radiotherapy, despite its significant properties, cannot stimulate the immune system, lonely. And in recent decades, researchers have found the combination of radiotherapy and immunotherapy as a more effective way to treat cancer.

4.3.2 Radioimmune therapy

During the past decade, attention to RIT has been greatly increased as a treatment for cancer due to increase abscopal effect. And this therapeutic approach is divided into several subgroups and each of them is different in the type of immunotherapy. In combination with radiotherapy for clinical and preclinical treatment of cancer some cytokines, transfer of some immune cells to the tumor environment, gene-mediated cytotoxic immune therapy (GMCI), vaccine therapy and monoclonal antibodies as some methods of immunotherapy are used [32,28]. For example, in 2012, Sampson et al. have proven that combination between RT and vaccination and a chimeric anti-IL-2 receptor monoclonal antibody (basiliximab) can inhibit the growth of the primary tumor in glioblastoma, and the tumor cell which have metastasized to lung [33]. Fig. 4.7 shows four main categories of RIT. In the next section, clinical and preclinical studies performed by researchers in each subgroup have been reviewed.

4.3.3 Clinical and preclinical studies of RIT for cancer treatment

One of the RIT's subgroups is transfer of immune cells to the tumor environment in combination with radio therapy in different doses. Injection of immune cells in the tumor enhances the effect of abscopal, as well as the complete elimination of tumor cells whose DNA is degraded by radiation therapy. In this regard Raj et al. has reported a study of neoadjuvant radiation with DC injections on patients with soft tissue sarcomas (STS). This treatment method has been composed of 50 Gy external beam radiation (EBRT) which has been applied in 25 fractions in combination with four injections of DCs inside the tumor. According to their results, no toxicity or unexpected phenomenon has occurred, and also the percentage of survival and lack of relapse of the disease was very satisfactory [34].

The other subgroup is related to the use of some methods to increase T cell responses. For instance, increasing tumor antigen presentation can increase the T cell response. One of the preclinical studies of this subject can be found in the study of Dewan et al. which was conducted in animal model of mice. They have injected TSA mouse breast carcinoma cells into mice and after the tumor was created, they have used an agonist of Toll-like receptor-7 (TLR-7), in combination with three fraction of 8 Gy of RT, and also they have used low-dose cyclophosphamide before start of treatment with imiquimod and RT. Their results have shown that with respect to the effect of TLR on the activation of DC and further increase of T cell response, this therapeutic approach has prevented tumor growth. And it is important to note that cyclophosphamide has increased the therapeutic effect [35].

The two following methods, vaccine therapy plus RT and the use of monoclonal antibodies in combination with RT, are important methods that have been considered in recent years. By introducing tumor antigens, the vaccine therapy stimulates the immune system against tumor antigens. In a study in 2016, an animal model of mice with melanoma was used to evaluate the efficacy of RT in combination with vaccine therapy. Combination treatment with focal RT and LM-based vaccine has been performed. They have used 16 Gy small-animal radiation research platform (SARRP) 7 days after tumor implantation. Then the results of this combined treatment with the result of using RT alone as a treatment or vaccine alone as treatment, have indicated that the use of RT as a treatment, minimal inflammation was seen in the treated tumor. In contrast, by using of LM-vaccination as a treatment, inflammation was increased. And the results of a treatment with combined method have shown presence of an intratumoral lymphocyte infiltrate and the results have indicated that the use of a combined method has led to an increase in the survival of this animal model with melanoma [36].

The use of antibodies to target antigens in addition to topical radiotherapy as well as the use of antibodies with other therapeutic methods, such as targeting a specific cytokine, for instance, TGFβ, to control a cancer treatment has been extremely attractive for researchers in the last decades. In this regard, Rodríguez-Ruiz et al. in 2019 has reported the use of monoclonal antibodies with radiotherapy has been shown to be effective in treating subcutaneous tumors of either 4T1 breast cancer cells or MC38 colorectal tumors induced in an animal model of mouse. By knowing that blocking TGFβ leads to an increase in abscopal effect, they have used anti-PD-1 and anti-CD137 monoclonal antibodies plus a TGFβ monoclonal antibody (1D11) with a regimen local radiotherapy and have managed to find out that the use of this method has increased the effect of abscopal [37].

4.4 Chemo immunotherapy

As the success rate of antibodies from the earliest humanized forms have not been less than 15%.

4.4.1 Chemotherapy

Radiotherapy and surgery are mainly used to eliminate solid tumors or normal tumors that are located in a particular area. In other word, radiotherapy and surgery are the therapeutic methods for locoregional diseases. However, these local therapies are not enough to treat metastatic cancers which involve cancerous cells entering the blood or lymph, and spreading in different parts of the body or hematologic malignancies. Therefore, systemic treatments are introduced. Chemotherapy is a treatment for advanced cancers that enter to the blood and lymph nodes, and it is also the main treatment for hematological cancers, such as leukemia. The mechanism of chemotherapy drugs for treating cancer is in the field of mitosis—cell division. More specifically, most chemotherapy agents damage cells with higher division rates, and since malignant and tumor cells grow faster than other healthy tissue cells, chemotherapy has been seen as the main treatment for cancer. But in addition to destroying and damaging cancer cells, chemotherapy also affects other healthy and normal cells which are fast growing. For instance, some of the cells in the bone marrow and the cells of hair follicles are fast division cells, and thus chemotherapy affects them. Myelosuppression is one of the side effects of chemotherapy. Myelosuppression is the decrease in production of blood cells (leukocytes, erythrocytes, and thrombocytes) [38].

4.4.1.1 Mechanism of action

Chemotherapy agents are known to be cytotoxic as they damage the cells, and the mechanism of their behavior is different (Fig. 4.8). Chemotherapy agents mainly cause apoptosis in cells with rapid cell division rates. These cytotoxic substances induce apoptosis by destroying DNA or preventing mitotic division [39].

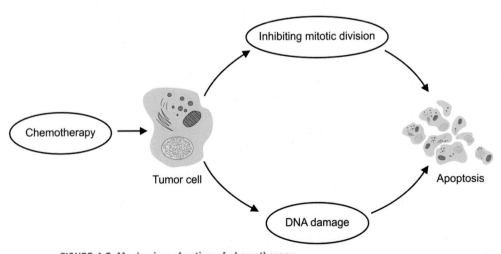

FIGURE 4.8 Mechanism of action of chemotherapy.

4.4.1.2   Chemotherapy regimens and drugs for the treatment of various cancers

Among the many strategies and chemotherapy drugs that are licensed by the FDA (Food and Drug Administration), the therapeutic regimen is selected based on the type of cancer and its progression. General guidelines have been obtained during clinical studies and clinical trials based on these two criteria. However, the choice of a proper chemotherapy regimen from various guidelines for cancer depends on several other factors in relation to the patient. These factors include age, history of other illnesses in the patient, previous treatment(s), hematological examination, liver and kidney function [40]. Table 4.1 shows examples of FDA-approved drugs that have been used in recent years to treat various cancers in clinical trial.

4.4.1.3   Chemotherapy, definitive treatment for cancer?

Despite the usefulness and importance of chemotherapy, the long history of its use, as well as the existence of different drugs and methods of treatment, chemotherapy is ineffective in destroying cancerous cells that do not have rapid cell division rates. And the known side effects of chemotherapy create barriers against the effectiveness of it on the destruction of tumor cells. These include the effects of it on healthy tissue and the normal cells, and the increasing resistance of tumor cells to chemotherapy drugs. A side effect of chemotherapy is the destruction of healthy cells that have high division rates. In Section 4.4.1, these effects are briefly cited. As mentioned above, one of the most prominent reasons for chemotherapy failure is the resistance of tumor cells to chemotherapy agents. The resistance of cancerous cells to chemotherapy drugs is mainly due to two reasons:

1. Some of the cancerous cells have been resistant to the drug from the start of treatment. That is, tumor cells have been resistant to the drug already.
2. The tumor cells were initially susceptible to the drug, but became resistant to it during the procedure of the treatment [40,41].

Another cause of chemotherapy failure as a definitive treatment is the presence of cancer stem cells and their resistance to chemotherapy. This problem, which has been

Table 4.1   FDA-approved drugs that have been used to treat various cancers in clinical trial.

Agent	Type of cancer	Reference
Bendamustine	Advanced chronic lymphocytic leukaemia (CLL10)	[41]
Docetaxel	Nonsmall-cell lung cancer (OAK)	[42]
FOLFOXIRI	Metastatic colorectal cancer	[43]
Paclitaxel	Metastatic breast cancer	[44]
Cyclophosphamide	Primary Central Nervous System Lymphoma (PCNSL)	[45]
Abiraterone acetate	Metastatic castration-resistant prostate cancer (COU-AA-302)	[46]

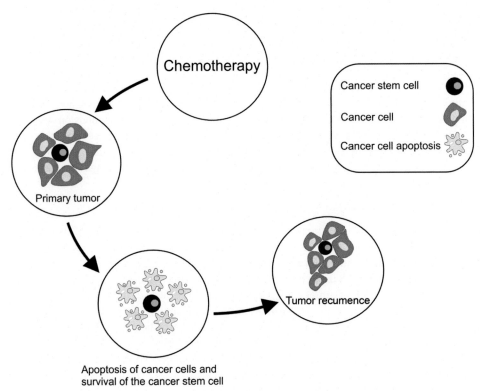

FIGURE 4.9 A schematic of a tumor containing a cancer stem cell.

less noticeable in the past, may be considered as the main issue of chemotherapy for cancer treatment now. Cancer stem cells are important due to the recurrence of disease and metastasis to other parts of the body [47]. In Fig. 4.9, a schematic representation of the cancer stem cell shows that, unlike cancer cells, it has not been apoptotic and has caused disease recurrence and the formation of a tumor in another region.

In Fig. 4.9, representation of a tumor containing a cancer stem cell that did not disappear after chemotherapy in spite of apoptosis of cancer cells. The surviving cancer stem cell after chemotherapy has caused a new tumor and recurrence of the disease.

Chemotherapy is commonly used in combination with other conventional treatments such as surgery, radiotherapy, hormone therapy and in recent years, immunotherapy.

4.4.2 Chemo immunotherapy

As mentioned in the previous section, chemotherapy is usually used in combination with radiotherapy, surgery and hormone therapy to treat cancer. But in recent years, it has been proven that these therapies are not sufficient for the definitive treatment of patients with cancer, and the possibility of a relapse of the disease exists even after using these therapies. However, the chemotherapy regimens which have been used with low doses of drugs reduce the number of T_{reg} cells which cause the formation of an immune environment in the tumor and will ultimately lead to the cancer cells death. However, the use of high-dose regimens which are recommended for some advanced cancers is immunosuppressive and will inevitably lead to the ineffectiveness of chemotherapy. Therefore, the use of low-dose chemotherapy regimens can be effective in stimulating the immune system to fight against tumor cells. However, it is not possible to use low-dose regimens in all cases, because of the progression of the disease. Additionally, the stimulatory effect of low-dose chemotherapy drugs on the immune system may not be enough in some cases. Therefore, the use of chemotherapy in combination with immunotherapy as an effective therapeutic treatment for systemic malignancies has been considered. In fact, immunotherapy removes parts of the tumor that are resistant to chemotherapy and those that likely to cause metastasis [48].

4.4.2.1 Mechanisms of chemo immunotherapy

The main pathway of chemo immunotherapy that is more widely considered is the use of chemical agents conjugated to monoclonal antibodies. Through these antibodies, the chemical agents have been specifically introduced into cancerous cells, and not only do they prevent the side effects of chemotherapy, but the highest amount of the dose injected into cancer cells is also reached. The monoclonal antibodies and their performance are discussed in detail in the previous chapters. The use of small molecular inhibitors is another mechanism of chemotherapy which work by blocking some signaling pathways, preventing abnormal cells from being removed by the physiological system of the body, eliminates abnormal cell proliferation and prevents the disease from reversing.

In general, chemo immunotherapy treatment against cancer occurs in two ways:

1. Causing the apoptosis of primary tumor cells by chemical agents that specifically target cancer cells, and then introducing tumor cell antigens to the immune system to prevent recurrence of the disease.
2. Destruction of tumor cells caused by chemical agents and changes in the signaling pathways of cancer cells to prevent recurrence of the disease [49].
To date, many chemo immunotherapy regimens have been investigated and clinically examined. One of the most famous chemo immunotherapies is CHOP (Cyclophosphamide, Hydroxydaunorubicin, Oncovin, and Prednisone) combined with RITUXIMAB, and this is used to treat B-cell non-Hodgkin lymphomas [50].

4.4.3 Clinical examination of chemo immunotherapy

In 2016, Fischer et al. has compared two therapeutic methods for the treatment of chronic lymphocytic leukemia (CLL) in a clinical study. They compared the Fludarabine and Cyclophosphamide (FC) regimen with Fludarabine, Cyclophosphamide and Rituximab (FCR) regimen on patients with CLL, and have shown that FCR was significantly more successful through evaluating the results of progression-free survival (PFS), median overall survival (OS) and the presence or absence of secondary malignancies in the long-term follow up [51].

In a phase II clinical study, Wang et al. has successfully managed using of chemo immunotherapy to receive good results for Mantle Cell Lymphoma treatment. In this clinical study, a relatively high chemo immunotherapy regimen with Retuximab with Hyper-CVAD (Hyper-CVAD chemotherapy consist of combination of drugs) was used [52].

Knutson et al. was also conducted a clinical study in 2016 on 48 women with metastatic HER2$^+$ breast cancer. They have reported that adding Trastuzumab to the chemotherapy regimen increases the survival rate of patients with HER2$^+$ breast cancer. In this study, patients received Paclitaxel, Carboplatin, and Trastuzumab. Their results have shown that the use of combination therapy (in here chemo immunotherapy) has been effective in inducing an adaptive immune system against antigens released by tumor cells [53].

In another study by Spanish researchers in 2019, the effect of Nintedanib, Docetaxel, anti-PD-1, and anti-PD-L1 has been proven to be a therapeutic regimen for the treatment of advanced lung adenocarcinoma [54].

4.5   Antibody-drug conjugate

The mid-1990s saw antibodies progress tremendously as drugs, with over 60 drugs produced based on antibodies. Antibodies are one of the most developed class of drugs that have a high potency in cancer treatments, autoimmune diseases and chronic inflammations [55]. In this chapter, the benefits of antibodies that are conjugated with drugs have discussed.

Problems have always existed in the traditional method of dealing with cancer. These methods have mostly relied on chemotherapy causing systemic cytotoxicity. The first inklings in monoclonal antibody production was altered the cancer therapy methods into new approaches, which was selective treatment [56].

ADCs are considered to be useful in cancer treatment, due to their capabilities in monoclonal antibody specification, the growth in linker fabrication strategies and possibility of toxic drug usage [57]. Antibodies in ADCs consist of chimeric (i.e., a combination of genetic materials from other species such as rabbit, mice, etc.) or humanized antibodies that become attached to the toxic drug by linker and cause local release of payload inside of antigen positive cells. Hence, the select ability of antibody and targeted release makes ADCs production as an interesting topic [58]. In Fig. 4.10, a schematic diagram reflects the main idea of the topic.

In 1913, the first steps of delivering a cytotoxic drug into cancerous cells with the usage of targeting materials was invented by Ehrlich [59]. Afterwards in 1970, antibodies were armed with a toxic agent as a means to kill specific cells. In 2000, the first commercialized ADC (gemtuzumab ozogamicin) against acute myeloid leukemia (AML) was approved, however in 2010 this ADC showed a high rate of

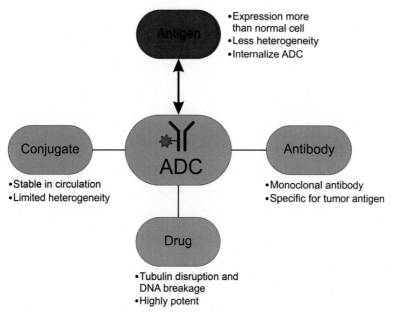

FIGURE 4.10 Antibody-drug conjugate.

mortality, so their production voluntarily stopped. Latest studies have revealed that a lower dosage of this ADC could be beneficial in fighting against AML [55].

More recently, three ADC drug have been approved. These are T-DM1, brentuximab vedotin, and inotuzumab ozogamycin, against CD22 which was approved in 2017 for those who suffer from ALL (acute lymphoblastic leukemia) and phase I and phase II clinical trials display promising results. Moreover, the combination of ADCs with other common treatments such as chemotherapy would increase the success rate [58,60].

In order to determine the ADCs pros and cons, the basis function of ADCs must be correctly understood. It would be worth mentioning that ADCs' selective ability leads to their movement into the tumor's location. Secondly, ADC attaches to cells that present antigen and are entered into endosomes/lysosomes. Finally, the toxic drug is released and links to its final target which would be: DNA, tubulin causing apoptosis. In some cases the toxic drug diffuses from dead cells, resulting in other abnormal or normal neighbor cell apoptosis and this happens due to a membrane permeability [61] named Bystander killing. This process has been shown in Fig. 4.11.

In Fig. 4.11, a schematic illustration of ADCs function by encountering the tumor cell is portrayed.

We can see the following:

a) ADCs movement to tumor environment
b) ADC binding to target antigen on the cell surface
c) Endocytosis of both antigen and ADC
d) Nonlysosomal drug released in cleavable payloads after ADC internalization

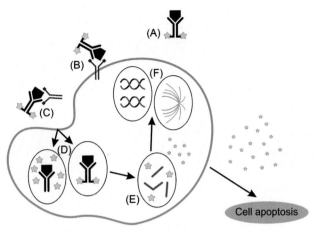

FIGURE 4.11 Bystander killing.

e) Lysosomal degradation of ADC in noncleavable payloads
f) Release of payloads causing disruption of microtubule or DNA break and finally
g) Cell apoptosis

Note: permeable membrane causing bystander killing in antigen negative cells as a result of drug release in its micro environment.

Although ADCs have the ability to expand the therapeutic window through its ability in antigen selection and usage of toxic agents [58,60,61], as well as increasing the safety by decreasing the toxic agent exposure with normal tissues, scientists are faced with obstacles and failures due to the usage of weak chemotherapy agents, linkers with instability and problems with antigen recognition, slow rates of antibody penetration from plasma to target tissue which causes low amount of ADCs to reach tumors. Further investigation revealed that after 24 h of injection, only 0.01% of administrated dose was reached to tumor cells [60-62].

The vital factors in ADCs manufacturing include: decreasing the immunogenicity, expanding the circulation span, boosting the antigen targeting specificity, and diminishing the normal cell injuries [63].

To satisfy all these factors it is essential to study the chemical properties of the linkers, the payload and antibody, the tumor antigen selection and the bio properties of the target cell surface [62,64]. ADCs are composed of three divergent parts: the antibody, the cytotoxic part (drug) and the linker. All these parts and the antigen selection are illustrated in this chapter.

4.5.1 Antigens

Some crucial factors that lead to a better ADCs response include:

1) The targeted antigen should be expressed in a great amount with minimal heterogeneity, the nonhomogenous tumor antigen expression may cause the neighboring cell's death as well as systemic toxicity as a result of the bystander effect. This is to say that the part of membrane which is free of tumor antigens lead to the spreading of drugs after internalization. Moreover, the rate of expression must be more than normal tissues.

2) The manufactured antibody must be able to become internalized into the cell by cell receptors and after ADCs administration, the tumor antigen down regulation must not occur.

3) An optimal ADC should have a minimum release of the cytotoxic drug before reaching to the target cell for reducing the adverse effect [60].

One of the most vital issues in cancer treatment is the ability to target tumor-initiating cells (TICs), which are cancer stem cells. These cells are the main factors responsible for the growth of tumors, metastasis, etc., therefore targeting TICs is crucial, hence why we will discuss this further.

4.5.1.1   Specific antigens in cancer stem cells

In 1994, scientists found that only a rare amount of cancer cells have the features of malignancy and can be metastatic. These cells are named cancer stem cells [65]. It is hypothesized that cancer stem cells can survive from common treatments like chemotherapy and also have self-renewal abilities [64,66], so from the aforementioned data, we can conclude that targeting cancer stem cell antigens may lead to patient treatment. Although it has appeared that some molecules that are expressed on the membrane of TICs like CD47, CD44, CD133, CD33, etc., these markers are mostly expressed in both cancer stem cells and healthy stem cells. Therefore some of which that display a higher expression rate in tumor stem cells are considered for ADCs usage [66]. Although a vast majority of scientists are in agreement with the cancer stem cell theory and attempts in dealing with these cells were continued, nobody has been successful in this regard, so investigating the possible reasons in these preceding failures is essential:

1) The initial theories on cancer stem cells considered the number of these cells at about one in a million but it has been shown that the number of these cells may differ from case to case, and could reach the number of one stem cell per four cancer cell.

2) As cancer stem cell markers are not specific, or unique, and as there are no series of markers to use to identify and isolate cancer stem cells, scientists could not purify CSCs in order to study their properties.

3) The problem with cancer stem cell plasticity is really important. That is to say that differentiated tumor cells could become cancer stem cells to compensate the loss of cancer stem cells, therefore it is vital to eradicate all tumor cells not just CSCs [65].

4.5.2   Antibodies

Antibody is a protein that is shaped like a "Y," and it is produced in blood by B cells when the immune system becomes activated. It is the ability of antibodies in targeting unique antigens (antibodies binding site, allows them specific cell type selection) [59] that has caused researchers to keep them as novel cancer therapeutic methods [67]. Historically, IgG (immunoglobulin G) is known as the most used form in antibody construction. However, while IgG is still prominent, scientists have also attracted to other forms of antibodies; over 30 antibodies have considered for oncology usage differing in solid tumors and hematological tumors [55]. It is worth emphasizing that antibodies targeting solid tumors require a deep penetration in order to thoroughly reach the tumor, so the usage of antibody fragments instead of whole antibody is recommended.

There are many different forms of antibodies with different uses, including those conjugated with drugs, radioisotopes (like heavy metals), and PEG (polyethylene glycol) or antibody fragments [55] and monoclonal antibodies, having been used in ADCs production. Due to the fact that antibodies must induce minimal antigenicity, monoclonal antibodies are mainly produced from humans or murine which is being humanized [66].

4.5.3 Linkers and conjugates

The function of linkers is to bond the payloads with the antibodies, and the physiochemical characteristics of linkers demonstrate the ADCs' pharmacokinetics [58]. In order to design a successful ADC, the linkers need to fulfill some special characteristics. First, linkers' stability in plasma is one of the crucial factors which helps them with better circulation and tumor targeting. Unstable linker causes drug leakage and systemic toxicity, resulting in damage to healthy cells. To produce stable linkers: payload, the type of cancer and antigens must be carefully studied. Despite stability in circulation, the linker should be capable of rapid drug release once ADC enters the tumor cells [62,63]. The final key criteria in designing the linkers are based on their hydrophobic bounds. Linkers with a hydrophobic manner along with hydrophobic drugs may trigger the accumulation of ADCs which have the ability in immune undesired alertness and liver removing effect. Hence, manufacturing hydrophilic linkers coupled with negative sulfonate, PEG and pyrophosphate diester groups would be demanding [63].

There are two general classes of linkers: cleavable and noncleavable. The former cleaves once ADC encounters the tumor cell and the latter group is not cleavable during the process [62].

- Cleavable

This group of linkers are produced to break when faced environmental changes. Hence, these linkers cleave when they move from an extracellular matrix into intracellular milieu by feeling the alternation in pH, redox potency (hydrolysis) or some determined enzymes (proteolysis), etc. once the linker breaks and the drug releases [58,63]. Those antibodies whose function is on the basis of proteolysis recognized by specific enzymes like cysteine proteases due to existing determined sites [58].

- Noncleavable

This class of linkers include stable and degradation-resistant bonds resulting in whole ADC degradation in the lysosome [63].

A huge gamut of ADCs is constructed by accidental connection of residues of lysine or cysteine to the antibody which resulted in ADCs with different DAR (drug-antibody ratios) [66]. Although many ADCs such as Mylotarg and Kadcyla (gemtuzumab ozogamicin and trastuzumab emtansine, respectively) are produced by lysine conjugation (this modification on the surface of antibodies is of the most common conjugation strategies), this method would provide heterogeneous mixture with divergent DAR leading to a low therapeutic window. Hence, novel ADC construction approaches have focused on site-specific strategies that provide ADCs with the exact DAR, better therapeutic windows, and many advances on developing less heterogeneous conjugation in lysine modification [66,67].

4.5.4 Drugs

Cytotoxic drugs are characterized in two groups: drugs which inhibit the microtubule and those which damage DNA. Early ADCs were included in classic chemotherapy payloads like methotrexate or doxorubicin due to the known toxic effects of these drugs, however their low instability/limited payload efficiency caused them to be ineffective. Then the second generation of ADCs came through, and the potency of the drugs used in this group was about 100–1000 folds higher [58,60].

4.5.4.1 *The most important drugs used in ADCs are discussed further*

Auristatins are the most common group of drugs in ADCs inhibiting tubulin gathering and lead to G2/M cell cycle discontinuing. Maytansinoids are the other microtubulin blockage drugs which are well recognized and both Kadcyla (trastuzumab emtansine) and Adcetris (brentuximab vedotin) are approved ADCs and fall into the Maytansinoids and Auristatins categories, respectively. One of the novel tubulin suppressor payloads which leads to microtubules instability and mitotic discontinuing is Tubulysins that forms stable ADC along with trastuzumab. It is worth noting that this group of drugs are efficient on cells with high proliferation rates such as most cancer cells.

The second group of drugs which damage the DNA are more effective in solid tumors. A number of famous DNA damaging ADCs include Duocarmycin and Calicheamicin [60,61]. It is conventional to add that these ADCs have the ability to kill cancer stem cells, however it also runs the risk of killing normal cells that express a low amount of tumor antigens [61]. Current ADC payloads are extremely toxic if they were freely used, as conjugation with antibodies reduce drug potency [61]. Mitotic inhibitor drugs prevent mitosis by chromosome division and the changing of the cytoskeleton structure as to promote cell death like Vinca alkaloids and taxoids [56]. Some prominent drugs in ADCs include: pertuzumab, rituximab, trastuzumab, and cetuximab [60].

4.5.5   Case study

In a study, Connors et al. have used brentuximab vedotin, which targets CD30, along with chemotherapy for those who suffer from Stage III or IV Hodgkin's Lymphoma. The results have revealed that the overall rate of death and cancer progression was 4.9% lower for patients who were treated with the combination of chemotherapy and ADCs, compared to those who treated with chemotherapy alone [68]. In another investigation in 2018, phase I clinical trial on the efficiency of vadastuximab talirine (which is an antibody) drug conjugate as single agent in AML patient with CD33-positive was tested. The results have illustrated the safety and activity of vadastux-imab talirine as monotherapy, and it is worth noting that the recommended dosage for this drug is 40 µg/kg [69].

4.5.6   Prospect

As discussed, the knowledge of ADCs manufacturing has been greatly developed in the past few decades and there are approximately 60 ADCs in clinical phase for both hematological and solid tumors, and the approval of three ADCs raise the hope of progressive usage of ADCs as novel cancer treatment methods in the near future. Some key factors in developing new ADCs include:

1. A growing demand in recognizing cancer-specific antigens
2. Producing ADCs which are stable in the circulation system and are less immunogenic
3. Inventing better payloads and methods to deliver more drugs into the target cells
4. Novel approaches in linkers and conjugation chemistries.

It is paramount to recall some problems scientists have faced in ADCs designing, such as: the expression of target antigens in normal cells, heterogeneity of both tumor cells and ADCs, low amount of payloads reaching the target antigen, instability of linkers causing systemic toxicity, survival of cancer stem cells—leading to relapse, etc.

Novel methods have solved many of the aforementioned obstacles. For example, site-specific conjugation resulting in near homogenous ADCs which increase the DAR and therapeutic windows, humanized antibodies in decreasing the immunogenicity to some extent, multiepitope targeting that helps scientists generalize ADCs to the vast majority of people, usage of antibody fragments instead of whole antibody to improve the ADCs penetration and finally, combination therapy which is of a great importance. The combination of chemotherapy along with ADCs would be recommended.

4.6   Cancer vaccine

The earliest indications of vaccines against tumors came into existence only after the invention of tumor-specific antigens (TSAs) in 1967, which were pioneered by George Klein. However, problems such as, low immunogenicity, toxicity and targeting issues was existed [70]. During the last decade, cancer therapies have significantly

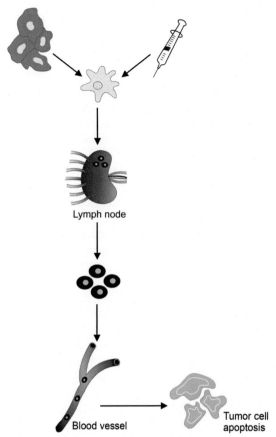

FIGURE 4.12 Schematic of cancer vaccine function or release of tumor cell antigen as a result of cancer cell death.

advanced. Moreover, immunotherapy has seen tremendous growth by demonstrating that T cells can show tumor suppression and antitumor activity. Tumor cells typically express mutated proteins or normal proteins in a higher rate which cause them to be recognized by CTLs (cytotoxic T lymphocytes) [71].

Schematic of cancer vaccine function (or release of tumor antigens as a result of tumor cell death) are shown in Fig. 4.12. Firstly, APCs (DCs, neutrophils, lymphatic endothelial cells, and macrophages) must recognize the antigen, then the T cells need to be activated by APCs, and in the last stage activated T helpers as well as T cytotoxic must move to the tumor site and change the immunosuppressive microenvironment to an inflammatory condition. This alteration in the tumor microenvironment helps the cytotoxic T cells to fight against T cells [70].

First, by vaccine administration or tumor antigen release APCs recognize tumor antigens and represent them on their surface, then they migrate into Lymph node and cause T cells differentiation and proliferation. Afterwards, mature T cells go through blood vessel and move into tumor cells environment. Finally, T cells lead to tumor cell apoptosis.

Immunotherapy methods include cancer vaccines, antibodies to destroy cancer cells, immune regulatory molecules, cell-based therapies [72].

Immunotherapy is an efficient method of combating tumor cells, and it can be used in combination with other common therapies such as chemotherapy and radiation therapy. Although some failures have been recorded in the history of cancer vaccinations, they are still considered to be one of the most highly developed fields in immunotherapy [73]. Most recently, personalized vaccines have provided scientists with new approaches in cancer vaccinations [74].

There are two different approaches in cancer vaccination: (1) prevent healthy people to get cancers (scientists have developed vaccines against HPV and Hepatitis B) and (2) treatment of patients with cancer (therapeutic vaccines stimulate the immune system to fight against cancer) [73].

The early stages of cancer immunotherapy, pioneers primarily focused on melanoma which leads them to achieve some noticeable results, as an example on melanoma: a phase I trial on melanoma lysates in combination with adjuvant (DETOX) administrated on 22 patients, and showed low toxic effects [75]. Over time, valuable improvements in cancer vaccine technology resulted in tremendous vaccines on the basis of DC vaccine, peptide/protein vaccine, viral vaccines, tumor cell vaccine as well as DNA and mRNA vaccines [70,73,76].

From 1995 to 2004, NCI Surgery Branch administrated over 500 vaccines for the treatment of approximately 440 patients who suffered from metastatic cancer, and the investigations were mostly centralized on the basis of peptide and viral vaccines which resulted in a major impact on those patients [77].

Although recent researches on cancer vaccination have reached to phase II and phase III clinical trial, the usage of therapeutic cancer vaccines did not represent clear usefulness in the patients who suffer from cancer and it may be useful in premalignant or low residual disease. Hence, combination therapy is of a great concern that can to say that, the combination of vaccines along with other methods, including immune checkpoint blockade, chemotherapy, radiotherapy and adoptive cell transferring have considered to be more efficient [78]. Fig. 4.13 represents the following discussions.

4.6.1   Main strategies in cancer vaccine production

Scientists focused on different methods to prepare cancer vaccines, there are five main strategies in cancer vaccine technology, including:

- Dendritic cell vaccine
- Peptide vaccine
- DNA/mRNA vaccines
- Tumor cell vaccine
- Viral and bacterial vaccine

Each strategy is discussed further.

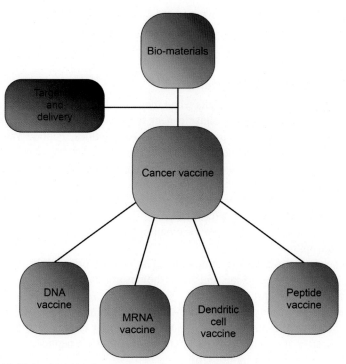

FIGURE 4.13 Main strategies in cancer vaccine production.

It would be worthy to illustrate divergent strategies in the way of cancer vaccine production. In spite of ancient infectious prohibition vaccines with the purpose of prevention, cancer vaccines are on the basis of triggering the immune system, prominently the CD8 + T cell in order to combat the cancer [73].

4.6.1.1 Dendritic cell vaccine

Dendritic cells control the activity of B cells and T cells, by moving and migrating to lymphatic organs and by secretion of cytokines, they ignite immune responses [79].

From the aforementioned information, DCs have considered to be one of the most useful APCs, that could help the sensitization of MHC restricted T cells to commence the immune responses [79,80]. To illustrate the functional ability of DCs in cancer vaccination it is not necessary to mention that tumor has the ability to suppress the maturity and trigger apoptosis of body DCs. DCs vaccines are exogenous APCs which could compensate the tumor suppression effect and motivate specific T cells recognizing tumor antigens [73].

In 1995 the first DC vaccine prepared for clinical trial on patients with metastatic melanoma, patients were administrated with APCs along with nanopeptide of MAGE-1 [81].

Some other investigation on DCs transfecting with tumor RNAs and DNAs, these vaccines have shown immunogenic responses [82-84].

Until 2003, more than 1000 DC vaccines were introduced and half of them shown minimal side effect [85]. These data reveal the safety of DCs usage for cancer vaccination.

4.6.1.2   Peptide/protein vaccine

The TAAs are recognized by T cells when they bind to MHC molecules. Then, activation of host immune system cause cancer tissue to be destroyed. Hence, cancer vaccines in this strategy are produced by TAAs peptides. Peptide vaccines have low toxicity profile, the production of these vaccines is somehow easy, and they are specific to their target and are not costly in comparison with other methods. Early peptide vaccines are prepared by single peptides causing problems, for example, at different stages of cancer, different peptides are expressed which is resulted in single peptides vaccines inefficiency [76,86].

The novel strategies in peptide vaccination have concentrated on targeting multiple peptide epitopes.

In order to distinguish the immunogenicity of multiple peptide vaccines, some clinical trials have been done and have shown promising result on patients with melanoma and glioblastoma [87,88]. The advantages of multiple epitopes have increased the probability of multiple T-cell activation and conquering the problem of loss or change of tumor epitopes (tumor peptides may change in different stage of cancers or can be different case by case) therefore, multiple peptide vaccinations can sufficiently keep T cell response [76]

More recently, scientists have found that some natural epitopes are not immunogenic. Therefore, some modifications are indeed vital to enhance their immunogenicity. Modified peptides are more capable to enhance the T cell immunogenity [73].

There are two different approaches in peptide vaccine production:

a. Self-antigens

Most of the peptide based cancer vaccines are produced with nonmutated self-antigens. These TAAs are expressed in both normal and cancerous cells, but sometimes these antigens overexpress in tumor cells. For example, cancer testis antigens (CTAs) express in normal and tumor cells, however, in normal tissues they express only in germline cells. From aforementioned data it is understood that using of self-antigens as a peptide vaccine causing central tolerance and prevent sufficient immune response against tumor cells, on the other hand however, powerful immune responses lead to autoimmune problems [76].

b. Neoantigens

In order to define neoantigens, these peptides are mutation-based antigens merely existed in cancerous cells with lower risk of autoimmune side effects or central tolerance. These antigens named TSAs and are generated from mutations (point mutations, translocations, insertions or deletions), viral proteins and posttranslational modifications (PTM) [76,89].

Neoantigens are defined into two groups: shared neoantigens and personalized neoantigens. Shared neoantigens are those which are common in patients with the same tumor type and not expressed in normal tissues. These antigens have the benefit of general vaccination for people suffering from similar cancers. However, neoantigens may be variable from cancer to cancer or in a similar cancer neoantigens are different in each patient. So, personalized vaccines seem to be the best vaccines which could give sufficient immune responses [76].

Personalized peptide vaccines are developed in order to solve the problem of peptide vaccines disability in triggering the immune system against cancer cells. In this method by consideration of the preimmunity status of patient, personalized cancer vaccine has been invented [73]. Some scientists believe that although this novel approach may boost the effect, such expensive methods with the low rate of publicity would cause further limitations.

4.6.2 DNA/mRNA vaccines

- DNA vaccines

The usage of DNA as a vaccine is illustrated in early 1990, plasmid DNA can promote immune system reaction against tumor antigens [90]. Plasmid DNA includes unmethylated and repeating cytosine-guanine area therefore, these DNA have the ability to trigger immunogenicity, so they play a role as both adjuvant and antigen [70,91]. There exist some problems with the usage of bare DNA, for example, miss-targeting and inefficient uptakes by the cells have happened in the previous researches. These troubles provide scientists with new efficient methods which cause the APCs to recognize plasmid DNA in a better way. Electroporation (EP) is a method by which the membrane of the cells become permeable with the help of electronic pulses, it has been revealed that this process extends the antigen uptake a lot. There seems to be one another fruitful way to boost the impact of the DNA vaccines. In this strategy the DNA is coated with metals (especially heavy metals like: gold), then cannonaded to the antibody presenting cells so, it will decrease the required dose adequately. Despite the aforementioned strives, only a few progression have been seen in DNA vaccination resulted from their low rate of immunogenicity [70].

It is worth nothing that, DNA vaccines seems to be efficient due to their simplicity, easy production and safety. Hence, there is ongoing researches on DNA vaccination especially in prostate cancer [92].

- mRNA vaccines

mRNA vaccines seem to be a good substitute for DNA vaccines because they are safer. In this approach, messenger RNA are used for immune alertness but it has been shown that mRNA vaccines are not stable leading to low development in these vaccines. More recently, with the help of stabilize improvement, mRNA vaccines get more attractive. These vaccines in combination with liposomes and protamines provide better immune reaction [73].

4.6.3 Tumor cell vaccines

In this method the TAAs of tumor cells are used for vaccination, these tumors could be autologous or allogenic. These tumors should become inactive before vaccination because live tumors may express immune-suppressive cytokines and also they are capable of producing new tumors in body. There are some methods to kill tumor cells, one of the most common strategies is the freeze-thaw method. By repetition of freeze and thaw process the tumor cell necrosis will happen and the cellular compartments including TAAs is released. Finally, TAAs are separated by centrifugation. Another used strategy is promoting tumor death by irradiation which causes tumor apoptosis and is milder than freeze-thaw method. Afterwards, the obtained TAAs must be loaded in biomaterial platforms which are discussed further.

The benefit of tumor cells as a source of antigens is the ability of tumor vaccines to produce strong immune response due to high amount of mutated antigens existing on cancer cells. Moreover, in the case of autologous tumor usage, the antigens become more personalized boosting the immune responses. However, some disadvantages of tumor vaccines is their high rate of cost and it cannot be administered to everyone in autologous tumor vaccines and low efficiency due to unspecific antigens in allogenic vaccines [70].

Novel approaches in tumor cell vaccination have revealed the advantages of using tumor cell lysate along with DCs. Different strategy on tumor cell lysate have newly explored, like vaccines with three-dimensional polymer matrices in combination with tumor lysates along with adjuvants and the efficiency of the system is proved in models of murine with melanoma [93].

4.6.4 Viral vaccines

In 1890s the usage of bacteria in cancer treatment has figured out. Coley accidentally has found that a patient suffering Sarcoma was getting better with erysipelas. These results caused him to try treatments with inoculation of *Streptococcus* and *Bacillus* prodigious. Thereby, he has declared that it was hard to induce meaningful infection with the usage of inoculation strategy. Since then, researchers have tested divergent bacteria including: *Listeria monocytogenes* (Lm), *Clostridium novyi*, and *Salmonella typhimurium*. Some of these organisms have been used as pure vectors and some of them in combination with adjuvant or DCs to express tumor antigens [89].

Viruses are other efficient organisms in cancer treatment. The usage of viruses is not limited to attenuated viruses in combating viral diseases by induction of protective immunity. With the help of genetic advances, viruses have used as vectors to express and deliver antigens in order to induce immune reaction, and this immune reaction is not only usable in viral disease, but also in cancerous cells which are not naturally immunogenic [89].

Recent attentions in utilizing viruses is not only limited to their usage as gene vectors, but as direct toxic candidates to tumor cells, these viruses are called OVs [94].

OVs are viruses that have the ability to infect and kill tumor cells [95]. Recent developments in the usage of these OVs have resulted in a huge realm of engineered viruses with immune alerting ability. These viruses have become famous in cancer vaccination due to their direct cytolytic effect on tumor cells. OVs are interesting candidates in cancer immunology as they can induce immune responses against cancer cells [93]. Adenovirus is the most researched OV [94].

Although, adenoviruses (Ad2 and Ad5 are the most common ones) are capable of TAAs delivering, a lot of people are shown low immunity against adenovirus due to long time exposure and hence produced two similar systems [89].

One of the concerns about these vaccines is that, the usage of attenuated live viruses in patients who have recently taken chemotherapy may cause frank infection or other health problems [94].

In one study, scientists have produced two vaccines: (1) with two different viruses and (2) combination of virus and bacteria. They have used combination of virus and bacteria in order to compare the results. They have produced two similar systems, one of them with adenovirus (Ad) and Maraba (MRB) and the other one with Lm which is a bacteria and Maraba (MRB) which is virus. Both vaccines administrated in mice models. Although both vaccines have shown sufficient immune response, the mice have shown smaller tumor size and longer survival in bacterial-viral system in contrast to the usage of different two viruses [96].

4.6.5   Biomaterials in delivery and targeting

One of the most crucial factors in cancer vaccination is targeting, and over the past decade scientists have been faced with problems of insufficient delivery. Hence, some new delivery systems have been invented and they have shown promising results. Different kinds of targeting and delivery systems including nano- and microparticles, scaffolds, and self-assembling materials have been prepared and used with different cancer vaccine strategies [70]. In this chapter two vital delivery system have been discussed:

- Cancer vaccines on the basis of nanoparticles

Targeting systems are made of polymers and lipids, in the following paragraph most famous nanoparticles have been discussed:

Liposomes considered to be one of the first used nano materials for cancer vaccination, that is, resulted from the FDA approval of these materials and their delivery capability. These two reasons causing liposomes as a promising candidate for vaccine delivery. On the other hand, some drawbacks in their loading capability, stability and toxicity limited their usage. PLGA is the other crucial nanoparticle for vaccine delivery. Although PLGA can be manufactured easily and can be prepared in different particle sizes, however, the low (drug) loading amount is one of the ongoing problems. In order to compare these two most used nanoparticles, it is worth mention that liposomes contain both hydrophobic and hydrophilic bonds which lead them to be suitable for carrying both hydrophobic and hydrophilic components at the same place. However, PLGA mostly contains of hydrophilic bonds results in a low hydrophobic loading capacity. To overcome the aforementioned obstacles scientists have worked on modification strategies to stabilize and enhance the loading rate of both Liposomes and PLGA [70].

- Cancer vaccines on basis of scaffolds

The reason why scientists are motivated to use scaffolds is due to the benefits of local delivery systems. Local delivery may decrease the systemic toxicity and boost the immune effect with lower doses [97].

Scaffold delivery systems mainly have made of polymers and hydrogels. These systems are typically large, so they remain in the injection site. Moreover, scaffold base delivery systems are capable in encapsulation of different molecules.

There are three efficient materials that have been used as cancer vaccines scaffolds, including PLGA, hydrogels which mainly contains of alginate and mesoporous silica micro-rods (MSRs), these scaffolds are all biodegradable and biocompatible [70].

The scaffolds play a key role in mimicking the natural lymph node function, so they are vital for immune cells to interact and expand. Based on these points, some factors in scaffold fabrication seems demanding, like: porosity and interconnection between porous which are essential for cell nutrition and cell encapsulation respectively, hence scaffolds must meet some factors, including: rigidity to bear the forces, porosity to provide convenient penetration, and control the signaling maters [97].

In conclusion, combination of biomaterials with cancer vaccines has shown promising results by reducing the dosage, increasing the safety due to their localized administration, boosting the efficiency resulted from their targeting ability and also biomaterials can influence on signaling. However, some obstacles exist in their usage: First, there are far differences between academic researches and clinical trials, animals may be treated in a different way from humans when they faced to biomaterials. Second, most of the investigations on biomaterials for cancer vaccination approaches have been done on melanoma which is different from solid and hematological tumors, so implantation seems to be different from cancer to cancer. Finally, some troubles in commercial and clinical purposes have been happened, including: the large scale production, and FDA approval. In order to facilitate the clinical trial phases it is better to use FDA approved biomaterials like PLGA [70,71].

It is worth noting that, on one side, designing suitable preclinical platforms for cancer vaccines is difficult due to the differences in human and mouse immune systems. On the other hand, mouse tumors may not always mimic human cancers in reality [89].

4.6.6 Case study

Maslak et al. have succeeded to produce a vaccine named galinpepimut-S against WT1 protein as a treatment for those who suffer from AML. In phase II study patients were administered vaccines for at least six doses in 10 weeks. The results have revealed that the vaccine is well tolerated and 22 patients were well treated [98].

In a clinical trial phase1/1b study the OTSGC-A24 cancer vaccine was used in 24 patients with gastric cancer for 4 weeks in order to evaluate the safety and immune response on the basis of induction of CTL. The results have demonstrated the safety of the vaccine and 15 patients showed positive CTL after 4 weeks [99].

4.6.7 Future prospect

It would be clear that combination of cancer vaccine therapy with other methods including: radiotherapy and chemotherapy boost the results, so this vital issue is of great concern. Moreover, one of the most important complications in cancer treatment is the vitality of cancer stem cells, hence developing vaccines with the ability of stem cell recognition and killing is demanding. Furthermore, there are some well-known stem cell markers such as Vimentin [100], it is important to understand whether they express on the surface of the cell or inside the cells in tumor cells, whenever we find the markers location, we can better manage our vaccines. As a last point it is worth mentioning that, although newly manufactured vaccines are focused on personalized vaccination, finding global and general methods for cancer vaccination with the usage of multiple fragments would be demanding.

4.7 Nonspecific immunotherapy

In general, the goal of all cancer therapeutic is the eradication of cancer cell colony and to prevent metastasis [101]. However, nonspecific Immunotherapy has focused less on identification and destruction of the malignant cell, so it has more attention on inducing widespread changes between immune system and cancer cells in order to make a modulation on this battle. These therapies are less sensitive to antigen-specific therapies. It is also known as immunomodulatory therapies [100,101]. Although it mostly has used as an adjuvant (along with the main treatment) to improve the immune system, another type of immunotherapy like a vaccine, however, some nonspecific immunotherapies have given themselves as a cancer treatment. These therapies can include cytokines and chemicals released by immune system cells such as interleukins and interferons which are essential for growth and controlling the immune system cells and blood cells. Usage of drugs like immune inhibitor checkpoints which are similar with cytokines can boost the immune system, but unlike cytokines, these are not naturally found in the body so they are injected to the body like a drug [101-103].

Nonspecific immunotherapy has different mechanisms that can be categorized into direct antitumor effects, activation of innate immunity, a reversal of immune suppression and antigen-nonspecific T-cell activation. And they likely include macrophages, non-MHC-restricted killing by NK cells and memory T cells. They are varied to each other and some of them are too complex which have not been understood until recent years. Fig. 4.14 clearly shows the mechanism of nonspecific immune therapy via specific immune therapy [101].

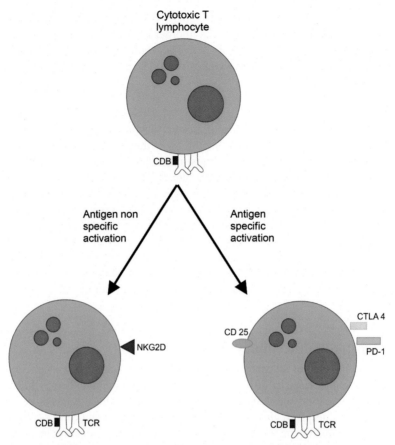

FIGURE 4.14 The mechanism of nonspecific immune therapy via specific immune therapy.

4.7.1 Cytokines in nonspecific immunotherapy

Cytokines have known as molecular messengers and cells of innate and adaptive immunity in response to exterior agents such as microbes or tumor antigens produce these molecules. These molecules allow the immune system's cells to communicate to each other. The secretion of cytokines is different from other immune system communication. Generally, another communication occurs through direct cell-cell interaction, while in cytokines it happens by the propagation of multifaceted immune signaling [101,104]. It has proven that cytokines have broad antitumor activity in cancer therapy which led to the eradication of cancer cells. And also this signaling is characterized by a significant amount of pleiotropic, in which one cytokine has the ability to act on many different cell types to mediate diverse and sometimes opposing effects. They have an important role to control the growth and activity of immune system cells. This role depends on several factors, including the local cytokine concentration or the pattern of a cytokine receptor. In cancer immunotherapy sometimes there is a need to inject the cytokines in order to amplify tumor-specific immune responses near to other treatments like a vaccine to produce a highly efficient effect. The most common ones are discussed here.

4.7.1.1   Interleukins

Interleukins is a cytokine group that acts in signaling between cells such as white blood cells and regulate cell growth, differentiation, and motility [104]. Each interleukin has its own role and they have a different effect on the immune system. Studies on animal tumor model have demonstrated that some cytokines especially interleukins have a broad antitumor activity which this feature has made cytokine-based approaches for cancer therapy. In 2015, Tugues et al. have demonstrated the therapeutic effects of IL-12 and have highlighted the advantage of the antitumor activity of IL-12 with its limited adverse effects that make this interleukin as a potent cytokine in mediating antitumor activity in a variety of preclinical models [105]. Other interleukins like GM-CSF, IL-7, IL-15, IL-18, and IL-21 have been entered clinical trials for patients with advanced cancer but most of them are used as an adjuvant not a single treatment. To date, Interleukin-2 (IL-2) has got FDA approval as a single agent for cancer treatment.

4.7.1.2   Interleukin-2

In 1980 IL-2 because of its ability to expand T-cells and sustaining T cell responses has been known as a powerful immune growth factor. IL-2 expands T cells without loss of functionality. This effect has caused the cytokine to be used in cancer immunotherapy. While immune-related adverse effect has been reported, but it has been demonstrated that IL-2 leads to complete and curative regression in the patients with cancer [106].

IL-2 also plays an important role in growth and activation factor for natural killer (NK) cells. IL-2 has a role in T cell proliferation, differentiation and activation. IL-2 with anti-IL-2 have a powerful signaling ability with the intermediate affinity CD122/CD132 receptor in vivo.

These promising outcomes make this cytokine as a single treatment in cancer immunotherapy. And now a man-made version of IL-2 [Aldesleukin (Proleukin)] has been produced in the laboratory and has been approved to treat advanced kidney cancer and metastatic melanoma.

IL-2 can be used as monotherapy. However, scientists have shown that high doses of IL-2 have adverse effect, hence it can be combined with another cytokine such as Interferon-Alfa or other cancers therapy like chemotherapy or radiotherapy. IL-2 also is used with other immunotherapy like adoptive cell transfer regimens, antigen-specific vaccination, and blockade of ICI molecules which can turn them more efficient against some cancers, but the side effects of the combined treatment are also increased.

IL-2 has showed a profound effect on cancer treatment. However, it has some side effects including chills, fever, and fatigue. Proleukin also has side effect such as gain weight and low blood pressure in some patients who have tried to treat with other medications. During high doses injection of IL-2 other side effects have been reported including heart beats and chest pain [106,107].

4.7.1.3   Interferons

In general, interferons (IFNs) are a chemical group of cytokines produced by host cells in response to the presence of pathogens, such as viruses, bacteria, parasites and tumor cells. They have a role to resist disease and infections. These group of proteins have an important effect on target cells by activating immune responses [108]. IFNs not only restraining the development of cancer but also it cause visible the tumor cells which escape from immune control. In addition, recent researches have demonstrated the antitumor effect of IFNs during the initiation or progression of several cancer types.

The types of IFN are named after the first three letters of the Greek alphabet:

- IFN-Alfa
- IFN-Beta
- IFN-Gamma

A recent study on a patient with breast cancer has shown that the cancer cells only responded to INF-Alfa. Hence, recently IFN-Alfa is used to treat cancer. It boosts the ability of certain immune cells to attack cancer cells. And also it has an effect on proliferation and differentiation of cancer cells and it has slowed the growth of cancer cells directly.

Platanias et al. have reported a strong effect of IFN-Alfa on the treatment of some malignancies including hairy cell leukemia (HCL) and chronic myelogenous leukemia (CML) [109]. IFN therapy has improved the rate of survival compared to other therapies like chemotherapy. IFN-Alfa also can be used to treat other cancers include follicular non-Hodgkin's lymphoma, kidney cancer, melanoma, cutaneous (skin) T-cell lymphoma, and Kaposi sarcoma [110].

INFs as a cancer treatment has some side effect such as:

- Flu-like symptoms (chills, fever, headache, fatigue, loss of appetite, nausea, and vomiting)
- Low white blood cell counts (which increase the risk of infection)
- Skin rashes
- Thinning hair

The appearance of these side effects is different in patients and most of them will disappear after stop of treatment, however, some of them such as fatigues can remains for a long time. Other side effects which are rare include nerve-damaging that mostly happens in the brain and spinal cord [108].

4.7.2   Immune checkpoint inhibitors

The goal of nonspecific immunotherapy is to boost the immune system without any specification. Beside the IL-2 and IFNs, there are other drugs which boost the immune system in a nonspecific way. But unlike cytokines, these are not naturally found in the body [111].

ICIs are proteins that block the cancer cells and stop the immune system of destroying normal cells. PD-1, PD-L1, and CTLA-4 are molecules that are targeted by ICIs [112].

After the approval of ipilimumab (anti-CTLA4) for the treatment of metastatic melanoma in 2011, five other immune checkpoints inhibitor got approval as an anti-PD-1 and PDL-1. The main function of these drugs is to remove inhibitory signals of T cells activation. But still, some point of their operation needs to be understood. One of this point is the expression of transforming growth factorβ (TGFβ) in cancers [113]. This factor drives immune dysfunction in the tumor because of inhibiting CD8+ and Th1. Ravi et al. have invented a biofunctional antibody-ligand trap which inhibits the expression of TGFβ and seeks to counteract T_{reg} cells. Their preclinical research has shown a positive result as compared to the usage of checkpoint inhibitor alone [114].

There are some other drugs such as Thalidomide, Lenalidomide, and Pomalidomide that are used in a nonspecific way. These drugs can cause some side effects including drowsiness, fatigue, and low blood cell counts.

Immunotherapy has been widely interested in recent years and researchers have found that using the body's immune system and stimulating it against disease is perhaps the best and only way to treat certain diseases such as cancer. In this regard, there are a variety of methods which have been tested in laboratories and in clinical experiences. On the other hand, eliminating cancer cells without damaging the normal cells and normal tissues of the patients in hematological malignancies or solid tumors has always been a wish for researchers. In this regard the advent of methods such as antibody based targeted therapy, scientists are getting closer to their dreams. However, traditional methods such as chemotherapy and radiotherapy that have been used to treat cancer in the past are still used in combination with immunotherapy, considering the benefits hereof and the positive effects it has had over the years on destroying cancer cells. But with all the benefits of chemo immune therapy and RIT, their effectiveness has been proven for the treatment of some cancers, and researchers are more concerned with methods such as cancer vaccination which extremely growing in the past decades. This method seems to be highly potent in cancer stem cell apoptosis. Furthermore, scientists have focused on personalized therapy which would be highly effective but with low publicity and globalization. Finally, ADC is expanding to new stages and site selective and multiepitope ADCs would boost the result and globalization. Hence, it is worth noting that although all of the discussed strategies have opened novel windows in cancer treatment, there exist tremendous obstacles forward, so combination therapy is still seeming to be the best approach.

In the following chapter another important cancer treatment method, cell therapy, will be discussed.

References

[1] J. Ferlay, I. Soerjomataram, R. Dikshit, S. Eser, C. Mathers, M. Rebelo, F. Bray, Cancer incidence and mortality worldwide: sources, methods and major patterns in GLOBO-CAN 2012, Int. J. Can. 136 (5) (2015) E359–E386.

[2] S. Sengupta, V.K. Balla. A review on the use of magnetic fields and ultrasound for non-invasive cancer treatment. *J. Adv. Res.* 14, (2018), 97–111

[3] C. Maccalli, K.I. Rasul, M. Elawad, & S. Ferrone. The role of cancer stem cells in the modulation of anti-tumor immune responses". In *Seminars in cancer biology*. Academic Press, 53, (2018), 189–200.

[4] D.M. Pardoll, The blockade of immune checkpoints in cancer immunotherapy, Nat. Rev. Can. 12 (4) (2012) 252–64.

[5] W. Manni, Y. Liu, Y. Cheng, W. Xiawei, & W. Yuquan. Immune checkpoint blockade and its combination therapy with small-molecule inhibitors for cancer treatment. *Biochimica et Biophysica Acta (BBA)-Reviews on Cancer*, 1871 (2), (2018), 199–224.

[6] Y. Wang, G. Marelli, A. Howells, N. Lemoine, Oncolytic viral therapy and the immune system: a double-edged sword against cancer, Front. Immunol. 9 (2018) 866–895.

[7] C. Yee, Adoptive T cell therapy: points to consider, Curr. Opin. Immunol. 51 (2018) 197–203.

[8] S.A. Rosenberg, N.P. Restifo, Adoptive cell transfers as personalized immunotherapy for human cancer, Science 348 (6230) (2015) 62–68.

[9] J. Scheller, E. Engelowski, J.M. Moll, & D.M. Floss. Immunoreceptor engineering and synthetic cytokine signaling for therapeutics. *Trends Immunol.*, 40(3), (2019), 258–272.

[10] W.J. Rettig, L.J Old, Immunogenetics of human cell surface differentiation, Annu Rev Immunol. 7 (1) (1989) 481–511.

[11] G. Köhler, C.J.n. Milstein, Continuous cultures of fused cells secreting antibody of pre-defined specificity, Nature 256 (5517) (1975) 495.

[12] D.G. Maloney, A.J. Grillo-López, C.A. White, D. Bodkin, R.J. Schilder, J.A. Neidhart, B.K.J.B. Dallaire, IDEC-C2B8 (Rituximab) anti-CD20 monoclonal antibody therapy in patients with relapsed low-grade non-Hodgkin's lymphoma, Blood 90 (6) (1997) 2188–2195.

[13] N.W. van de Donk, P. Moreau, T. Plesner, A. Palumbo, F. Gay, J.P. Laubach, P.J.B. Sonneveld, Clinical efficacy and management of monoclonal antibodies targeting CD38 and SLAMF7 in multiple myeloma, Blood 127 (6) (2016) 681–695.

[14] P.J.N.R.C. Carter, Improving the efficacy of antibody-based cancer therapies, Nat. Rev. Cancer. 1 (2) (2001) 118.

[15] A.M. Scott, J.D. Wolchok, L.J. Old, Antibody therapy of cancer, Nat. Rev. Cancer. 12 (4) (2012) 278.

[16] J. Patel, J. Amrutiya, P. Bhatt, A. Javia, M. Jain, A.J.J.o.m. Misra, Targeted delivery of monoclonal antibody conjugated docetaxel loaded PLGA nanoparticles into EGFR over-expressed lung tumour cells 35 (2) (2018) 204–217.

[17] C. Offenhäuser, F. Al-Ejeh, S. Puttick, K. Ensbey, Z. Bruce, P. Jamieson, A.J.C. Fuchs, EphA3 pay-loaded antibody therapeutics for the treatment of glioblastoma 10 (12) (2018) 519.

[18] R. Bazak, M. Houri, S. El Achy, S. Kamel, T.J. Refaat, C. Oncology, Cancer active targeting by nanoparticles: a comprehensive review of literature 141 (5) (2015) 769–784.

[19] A.L. Nelson. *Antibody fragments: hope and hype.* Paper presented at the MAbs, (2010).

[20] E. Grieger, G. Gresch, J. Niesen, M. Woitok, S. Barth, R. Fischer, C. Stein, Efficient targeting of CD13 on cancer cells by the immunotoxin scFv13–ETA′ and the bispecific scFv [13xds16]. 143 (11) (2017) 2159–2170.

[21] A. Nascimento, I.F. Pinto, V. Chu, M.R. Aires-Barros, J.P. Conde, A.M.J.S. Azevedo, P. Technology, Studies on the purification of antibody fragments. 195 (2018) 388–397.

[22] C. Mazzocco, G. Fracasso, C. Germain-Genevois, N. Dugot-Senant, M. Figini, M. Colombatti, F.J.S.r. Couillaud, In vivo imaging of prostate cancer using an anti-PSMA scFv fragment as a probe 6 (2016) 23314.

[23] D. Mercatelli, M. Bortolotti, A. Bazzocchi, A. Bolognesi, L.J.B. Polito, Immunoconjugates for osteosarcoma therapy: preclinical experiences and future perspectives. 6 (1) (2018) 19.

[24] N. Norton, N. Fox, C.-A. McCarl, K.S. Tenner, K. Ballman, C.L. Erskine, C.J.B.C.R. Calfa, Generation of HER2-specific antibody immunity during trastuzumab adjuvant therapy associates with reduced relapse in resected HER2 breast cancer 20 (1) (2018) 52.

[25] A. Chari, A. Suvannasankha, J.W. Fay, B. Arnulf, J.L. Kaufman, J.J. Ifthikharuddin, & R.J. B. Comenzo, R.J. B. Daratumumab plus pomalidomide and dexamethasone in relapsed and/or refractory multiple myeloma. blood-2017-2005-785246, (2017).

[26] R. Baskar, K.A. Lee, R. Yeo, K.-W.J. Yeoh, Cancer and radiation therapy: current advances and future directions. 9 (3) (2012) 193.

[27] D. Eriksson, T.J.T.B. Stigbrand, Radiation-induced cell death mechanisms. 31 (4) (2010) 363–372.

[28] S. Bockel, B. Durand, & E.J. C.R. Deutsch. Combining radiation therapy and cancer immune therapies: From preclinical findings to clinical applications, (2018).

[29] R.J.T.B.j.o.r. Mole, Whole body irradiation—radiobiology or medicine? 26 (305) (1953) 234–241.

[30] J. Krombach, R. Hennel, N. Brix, M. Orth, U. Schoetz, A. Ernst, S.J.O. Bierschenk, Priming anti-tumor immunity by radiotherapy: Dying tumor cell-derived DAMPs trigger endothelial cell activation and recruitment of myeloid cells. 8 (1) (2019) e1523097.

[31] R.R. Weichselbaum, H. Liang, L. Deng, Y.-X.J.N.r.C.o. Fu, Radiotherapy and immunotherapy: a beneficial liaison? 14 (6) (2017) 365.

[32] J.J.C.r.i.o.h. Bernier, Immuno-oncology: Allying forces of radio-and immuno-therapy to enhance cancer cell killing. 108 (2016) 97–108.

[33] J.H. Sampson, R.J. Schmittling, G.E. Archer, K.L. Congdon, S.K. Nair, E.A. Reap, I.I.J.E.J.P.o. Herndon, A pilot study of IL-2Rα blockade during lymphopenia depletes regulatory T-cells and correlates with enhanced immunity in patients with glioblastoma. 7 (2) (2012) e31046.

[34] S. Raj, M.M. Bui, G. Springett, A. Conley, S. Lavilla-Alonso, X. Zhao, & G.D. J.S. Letson, Long-term clinical responses of neoadjuvant dendritic cell infusions and radiation in soft tissue sarcoma, (2015).

[35] M.Z. Dewan, C. Vanpouille-Box, N. Kawashima, S. DiNapoli, J.S. Babb, S.C. Formenti, S.J.C.C.R. Demaria, Synergy of topical toll-like receptor 7 agonist with radiation and low-dose cyclophosphamide in a mouse model of cutaneous breast cancer. 18 (24) (2012) 6668–6678.

[36] C.M. Jackson, C.M. Kochel, C.J. Nirschl, N.M. Durham, J. Ruzevick, A. Alme, T.W.J.C.C.R. Dubensky, Systemic tolerance mediated by melanoma brain tumors is reversible by radiotherapy and vaccination. 22 (5) (2016) 1161–1172.

[37] M.E. Rodríguez-Ruiz, I. Rodríguez, L. Mayorga, T. Labiano, B. Barbes, I. Etxeberria, P. Berraondo, TGFβ blockade enhances radiotherapy abscopal efficacy effects in combination with anti-PD1 and anti-CD137 immunostimulatory monoclonal antibodies, Mol. Can. Therapeu. 18 (3) (2019) 621–631.

[38] M.V.J.L. Blagosklonny, Target for cancer therapy: proliferating cells or stem cells. 20 (3) (2006) 385.

[39] J. Zhao, Cancer stem cells and chemoresistance: The smartest survives the raid. *Phamacol. Therapeu. 160* (2016) 145–158.

[40] E. Chu, V.T. DeVita Jr., Physicians' Cancer Chemotherapy Drug Manual 2018, Jones & Bartlett Learning, (2017).

[41] B. Eichhorst, A.-M. Fink, J. Bahlo, R. Busch, G. Kovacs, C. Maurer, M.J.T.l.o. Sökler, First-line chemoimmunotherapy with bendamustine and rituximab versus fludarabine, cyclophosphamide, and rituximab in patients with advanced chronic lymphocytic leukaemia (CLL10): an international, open-label, randomised, phase 3, non-inferiority trial 17 (7) (2016) 928–942.

[42] A. Rittmeyer, F. Barlesi, D. Waterkamp, K. Park, F. Ciardiello, J. Von Pawel, M.C.J.T.L. Dols, Atezolizumab versus docetaxel in patients with previously treated non-small-cell lung cancer (OAK): a phase 3, open-label, multicentre randomised controlled trial 389 (10066) (2017) 255–265.

[43] C. Cremolini, F. Loupakis, C. Antoniotti, C. Lupi, E. Sensi, S. Lonardi, A.J.T.L.O. Zaniboni, FOLFOXIRI plus bevacizumab versus FOLFIRI plus bevacizumab as first-line treatment of patients with metastatic colorectal cancer: updated overall survival and molecular subgroup analyses of the open-label, phase 3 TRIBE study 16 (13) (2015) 1306–1315.

[44] V. Bernstein, S. Ellard, S. Dent, D. Tu, M. Mates, S. Dhesy-Thind, A randomized phase II study of weekly paclitaxel with or without pelareorep in patients with metastatic breast cancer: final analysis of Canadian Cancer Trials Group IND. 213. *167*(2), (2018), 485–493.

[45] Z. DeFilipp, S. Li, A. El-Jawahri, P. Armand, L. Nayak, N. Wang, Y.B.J.C. Chen, High-dose chemotherapy with thiotepa, busulfan, and cyclophosphamide and autologous stem cell transplantation for patients with primary central nervous system lymphoma in first complete remission 123 (16) (2017) 3073–3079.

[46] C.J. Ryan, M.R. Smith, K. Fizazi, F. Saad, P.F. Mulders, C.N. Sternberg, E.J.J.T.L.O. Small, Abiraterone acetate plus prednisone versus placebo plus prednisone in chemotherapy-naive men with metastatic castration-resistant prostate cancer (COU-AA-302): final overall survival analysis of a randomised, double-blind, placebo-controlled phase 3 study 16 (2) (2015) 152–160.

[47] M. Prieto-Vila, R.-u. Takahashi, W. Usuba, I. Kohama, T.J.I.j.o.m.s. Ochiya, Drug resistance driven by cancer stem cells and their niche. 18 (12) (2017) 2574.

[48] C. Da Silva, F. Rueda, C. Löwik, F. Ossendorp, L.J.J.B. Cruz, Combinatorial prospects of nano-targeted chemoimmunotherapy. 83 (2016) 308–320.

[49] E. Pérez-Herrero, A.J.E.j.o.p. Fernández-Medarde,& biopharmaceutics. Advanced targeted therapies in cancer: drug nanocarriers, the future of chemotherapy. *93*, (2015), 52–79.

[50] B. Coiffier, E. Lepage, J. Brière, R. Herbrecht, H. Tilly, R. Bouabdallah, P.J.N.E.J.o.M. Gaulard, CHOP chemotherapy plus rituximab compared with CHOP alone in elderly patients with diffuse large-B-cell lymphoma, New Engl. J. Med. 346 (4) (2002) 235–242.

[51] K. Fischer, J. Bahlo, A.M. Fink, V. Goede, C.D. Herling, P. Cramer, C.J.B. Maurer, Long-term remissions after FCR chemoimmunotherapy in previously untreated patients with CLL: updated results of the CLL8 trial, Blood 127 (2) (2016) 208–215.

[52] M. Wang, H.J. Lee, S. Thirumurthi, H.H. Chuang, F.B. Hagemeister, J.R. Westin, & W. Chen. Chemotherapy-free induction with ibrutinib-rituximab followed by shortened

cycles of chemo-immunotherapy consolidation in young, newly diagnosed mantle cell lymphoma patients: a phase II clinical trial. Am. Soc. Hematology, (2016).

[53] K.L. Knutson, R. Clynes, B. Shreeder, P. Yeramian, K.P. Kemp, K. Ballman, D.J.C.r. Northfelt, Improved survival of HER2+ breast cancer patients treated with trastuzumab and chemotherapy is associated with host antibody immunity against the HER2 intracellular domain, Cancer Res. 76 (13) (2016) 3702–3710.

[54] J. Corral, M. Majem, D. Rodríguez-Abreu, E. Carcereny, A. Cortes, M. Llorente, & T. Oncology. Efficacy of nintedanib and docetaxel in patients with advanced lung adenocarcinoma treated with first-line chemotherapy and second-line immunotherapy in the nintedanib NPU program. (2019), 1–10.

[55] P.J. Carter, G.A. Lazar, Next generation antibody drugs: pursuit of the'high-hanging fruit', Nat. Rev. Drug Dis. 17 (3) (2018) 197.

[56] R.V. Chari, M.L. Miller, W.C. Widdison, Antibody–drug conjugates: an emerging concept in cancer therapy, Angewandte Chemie Int. Ed. 53 (15) (2014) 3796–3827.

[57] S.C. Alley, N.M. Okeley, P.D. Senter, Antibody–drug conjugates: targeted drug delivery for cancer, Curr. Opin. Chem. Biol. 14 (4) (2010) 529–537.

[58] C. Chalouni, S. Doll, Fate of antibody-drug conjugates in cancer cells, J. Exp. Clin. Can. Res. 37 (1) (2018) 20.

[59] P. Ehrlich, Address in pathology, on chemiotherapy: delivered before the Seventeenth International Congress of Medicine, Brit. Med. J. 2 (2746) (1913) 353.

[60] N. Diamantis, U. Banerji, Antibody-drug conjugates—an emerging class of cancer treatment, Brit. J. Can. 114 (4) (2016) 362.

[61] P.A. Trail, G.M. Dubowchik, T.B. Lowinger, Antibody drug conjugates for treatment of breast cancer: novel targets and diverse approaches in ADC design, Pharma. Therap. 181 (2018) 126–142.

[62] J.M. Lambert, A. Berkenblit, Antibody–drug conjugates for cancer treatment, Ann. Rev. Med. 69 (2018) 191–207.

[63] K. Tsuchikama, Z. An, Antibody-drug conjugates: recent advances in conjugation and linker chemistries, Protein Cell 9 (1) (2018) 33–46.

[64] K. Starbuck, L. Al-Alem, D.A. Eavarone, S.F. Hernandez, C. Bellio, J.M. Prendergast, J. Behrens, Treatment of ovarian cancer by targeting the tumor stem cell-associated carbohydrate antigen, sialyl-thomsen-nouveau, Oncotarget 9 (33) (2018) 23289.

[65] K. Garber. Cancer stem cell pipeline flounders (2018).

[66] F. Marcucci, E.L. Romeo, C.A. Caserta, C. Rumio, Antibody-drug conjugates (ADC) against cancer stem-like cells (CSC)–is there still room for optimism? Front. Oncol. 9 (2019) 167.

[67] V. Chudasama, A. Maruani, S. Caddick, Recent advances in the construction of antibody–drug conjugates, Nat. Chem. 8 (2) (2016) 114.

[68] J.M. Connors, W. Jurczak, D.J. Straus, S.M. Ansell, W.S. Kim, A. Gallamini, E. Lech-Maranda, Brentuximab vedotin with chemotherapy for stage III or IV Hodgkin's lymphoma, New Engl. J. Med. 378 (4) (2018) 331–344.

[69] E.M. Stein, R.B. Walter, H.P. Erba, A.T. Fathi, A.S. Advani, J.E. Lancet, S. Faderl, A phase 1 trial of vadastuximab talirine as monotherapy in patients with CD33-positive acute myeloid leukemia, Blood 131 (4) (2018) 387–396.

[70] R. Zhang, M.M. Billingsley, M.J. Mitchell, Biomaterials for vaccine-based cancer immunotherapy, J. Control. Release 2 (2018) 256–276.

[71] S.T. Koshy, D.J. Mooney, Biomaterials for enhancing anti-cancer immunity, Curr. Opin. Biotech. 40 (2016) 1–8.

[72] K. Palucka, J. Banchereau, Dendritic-cell-based therapeutic cancer vaccines, Immunity 39 (1) (2013) 38–48.

[73] Q. Song, C.D. Zhang, X.H. Wu, Therapeutic cancer vaccines: From initial findings to prospects, Immunol. Lett. 196 (2018) 11–21.

[74] U. Sahin, E. Derhovanessian, M. Miller, B.P. Kloke, P. Simon, M. Löwer, T. Omokoko, Personalized RNA mutanome vaccines mobilize poly-specific therapeutic immunity against cancer, Nature 547 (7662) (2017) 222–226.

[75] M.S. Mitchell, J. Kan-Mitchell, R.A. Kempf, W. Harel, H. Shau, S. Lind, Active specific immunotherapy for melanoma: phase I trial of allogeneic lysates and a novel adjuvant, Can. Res. 48 (20) (1988) 5883–5893.

[76] A.R. Aldous, J.Z. Dong, Personalized neoantigen vaccines: A new approach to cancer immunotherapy, Bioorg. Med. Chem. 26 (10) (2018) 2842–2849.

[77] S.A. Rosenberg, J.C. Yang, N.P. Restifo, Cancer immunotherapy: moving beyond current vaccines, Nat. Med. 10 (9) (2004) 909.

[78] S.H. van der Burg, R. Arens, F. Ossendorp, T. van Hall, C.J. Melief, Vaccines for established cancer: overcoming the challenges posed by immune evasion, Nat. Rev. Can. 16 (4) (2016) 219–233.

[79] J. Banchereau, R.M. Steinman, Dendritic cells and the control of immunity, Nature 392 (6673) (1998) 245–252.

[80] R.M. Steinman, The dendritic cell system and its role in immunogenicity, Ann. Rev. Immunol. 9 (1) (1991) 271–296.

[81] B. Mukherji, N.G. Chakraborty, S. Yamasaki, T. Okino, H. Yamase, J.R. Sporn, J. Meehan, Induction of antigen-specific cytolytic T cells in situ in human melanoma by immunization with synthetic peptide-pulsed autologous antigen presenting cells, Proc. Nat. Acad. Sci. 92 (17) (1995) 8078–8082.

[82] A. Heiser, M.A. Maurice, D.R. Yancey, N.Z. Wu, P. Dahm, S.K. Pruitt, J. Vieweg, Induction of polyclonal prostate cancer-specific CTL using dendritic cells transfected with amplified tumor RNA, J. Immunol. 166 (5) (2001) 2953–2960.

[83] S.K. Nair, M. Morse, D. Boczkowski, R.I. Cumming, L. Vasovic, E. Gilboa, H.K. Lyerly, Induction of tumor-specific cytotoxic T lymphocytes in cancer patients by autologous tumor RNA-transfected dendritic cells, Ann. Surg. 235 (4) (2002) 540–556.

[84] C. Condon, S.C. Watkins, C.M. Celluzzi, K. Thompson, L.D. Falo, DNA–based immunization by in vivo transfection of dendritic cells, Nat. Med. 2 (10) (1996) 1122–1130.

[85] D. Ridgway, The first 1000 dendritic cell vaccinees, Can. Investig. 21 (6) (2003) 873–886.

[86] G. Parmiani, C. Castelli, P. Dalerba, R. Mortarini, L. Rivoltini, F.M. Marincola, A. Anichini, Cancer immunotherapy with peptide-based vaccines: what have we achieved? Where are we going? J. Nat. Can. Inst. 94 (11) (2002) 805–818.

[87] C.L. Slingluff, G.R. Petroni, K.A. Chianese-Bullock, M.E. Smolkin, S. Hibbitts, C. Murphy, J.W. Patterson, Immunologic and clinical outcomes of a randomized phase II trial of two multipeptide vaccines for melanoma in the adjuvant setting, Clin. Can. Res. 13 (21) (2007) 6386–6395.

[88] R. Rampling, S. Peoples, P.J. Mulholland, A. James, O. Al-Salihi, C.J. Twelves, J. Lindner, A cancer research UK first time in human phase I trial of IMA950 (novel multipeptide therapeutic vaccine) in patients with newly diagnosed glioblastoma, Clin. Can. Res. 22 (19) (2016) 4776–4785.

[89] H.M. Maeng, J.A. Berzofsky, Strategies for developing and optimizing cancer vaccines, F1000 Res. 8 (2019).

[90] M.A. Kutzler, D.B. Weiner, DNA vaccines: ready for prime time? Nat. Rev. Genet. 9 (10) (2008) 776.

[91] D.M. Klinman, G. Yamshchikov, Y. Ishigatsubo, Contribution of CpG motifs to the immunogenicity of DNA vaccines, J. Immunol. 158 (8) (1997) 3635–3639.

[92] C.D. Zahm, V.T. Colluru, D.G. McNeel, DNA vaccines for prostate cancer, Pharmacol. Therapeut. 174 (2017) 27–42.

[93] A. Aitken, D. Roy, M.-C. Bourgeois-Daigneault, Taking a stab at cancer; oncolytic virus-mediated anti-cancer vaccination strategies, Biomedicines 5 (1) (2017) 3.

[94] J.D. Denham, et al. Two cases of disseminated infection following live organism anti-cancer vaccine administration in cancer patients, Int. J. Infect. Dis. 72 (2018) 1–2.

[95] P.P. Peruzzi, E.A. Chiocca, Cancer immunotherapy: A vaccine from plant virus proteins, Nat. Nanotechnol. 11 (3) (2016) 214.

[96] A.S. Aitken, et al. Brief communication; a heterologous oncolytic bacteria-virus prime-boost approach for anticancer vaccination in mice, J. Immunother. (HagerstownMd. : 1997) 41 (3) (2018) 125.

[97] J. Weiden, J. Tel, C.G. Figdor, Synthetic immune niches for cancer immunotherapy, Nat. Rev. Immunol. 18 (3) (2018) 212.

[98] P.G. Maslak, T. Dao, Y. Bernal, S.M. Chanel, R. Zhang, M. Frattini, R. Rampal, Phase 2 trial of a multivalent WT1 peptide vaccine (galinpepimut-S) in acute myeloid leukemia, Blood Adv. 2 (3) (2018) 224–234.

[99] R. Sundar, S.Y. Rha, H. Yamaue, M. Katsuda, K. Kono, H.S. Kim, W.P. Yong, A phase I/Ib study of OTSGC-A24 combined peptide vaccine in advanced gastric cancer, BMC Can. 18 (1) (2018) 332.

[100] A.L.S. Satelli, *Vimentin in cancer and its potential as a molecular target for cancer therapy*, Cell. Mol. Life Sci. 68 (18) (2011) 3033–3046.

[101] A.M. Monjazeb, H.H. Hsiao, G.D. Sckisel, W.J. Murphy, The role of antigen-specific and non-specific immunotherapy in the treatment of cancer, J. Immunotoxicol. 9 (3) (2012) 248–258.

[102] S. Farkona, E.P. Diamandis, I.M. Blasutig, Cancer immunotherapy: the beginning of the end of cancer? BMC Med. 14 (1) (2016) 73.

[103] J.B. Swann, Y. Hayakawa, N. Zerafa, K.C. Sheehan, B. Scott, R.D. Schreiber, M.J. Smyth, Type I IFN contributes to NK cell homeostasis, activation, and antitumor function, J. Immunol. 178 (12) (2007) 7540–7549.

[104] O. Boyman, N. Arenas-Ramirez, Development of a novel class of interleukin-2 immunotherapies for metastatic cancer, Swiss Med. Week. 149 (0304.) (2019).

[105] S. Tugues, S.H. Burkhard, I. Ohs, M. Vrohlings, K. Nussbaum, J. Vom Berg, P. Kulig, B. Becher, New insights into IL-12-mediated tumor suppression, Cell Death Different. 22 (2) (2015) 237.

[106] T. Jiang, C. Zhou, S. Ren, Role of IL-2 in cancer immunotherapy, Oncoimmunology 5 (6) (2016) e1163462.

[107] S.A. Rosenberg, IL-2: the first effective immunotherapy for human cancer, J. Immunol. 192 (12) (2014) 5451–5458.

[108] B.S. Parker, J. Rautela, P.J. Hertzog, Antitumour actions of interferons: implications for cancer therapy, Nat. Rev. Can. 16 (3) (2016) 131.

[109] L.C. Platanias, Interferons and their antitumor properties, J. Interferon Cytokine Res. 33 (4) (2013) 143–144.

[110] B.L. Stein, R.V. Tiu, Biological rationale and clinical use of interferon in the classical BCR-ABL-negative myeloproliferative neoplasms, J. Interferon Cytokine Res. 33 (4) (2013) 145–153.

[111] A. Ribas, J.D. Wolchok, Cancer immunotherapy using checkpoint blockade, Science 359 (6382) (2018) 1350–1355.

[112] S.C. Wei, C.R. Duffy, J.P. Allison, Fundamental mechanisms of immune checkpoint blockade therapy, Can. Dis. 8 (9) (2018) 1069–1086.

[113] J.M. Michot, C. Bigenwald, S. Champiat, M. Collins, F. Carbonnel, S. Postel-Vinay, C. Massard, Immune-related adverse events with immune checkpoint blockade: a comprehensive review, Eur. J. Can. 54 (2016) 139–148.

[114] R. Ravi, K.A. Noonan, V. Pham, R. Bedi, A. Zhavoronkov, I.V. Ozerov, S. Nimmagadda, Bifunctional immune checkpoint-targeted antibody-ligand traps that simultaneously disable TGFβ enhance the efficacy of cancer immunotherapy, Nature Comm. 9 (1) (2018) 741.

Cell therapy

5

Chapter outline

5.1 Adoptive cell-based therapy in combination with chemotherapy

The effect of the immune system is to eliminate cancer cells before clinical expression. It provides the immunological reaction to the specific antigens of tumor cells that can eradicate them [1]. It happens when there is a balance between cancer and the immune system. Once this balance disturbed, tumor cells can develop and have clinical expression. Immunotherapy techniques are to boost the immune response against tumor cells, and they have shown themselves as the first broadly techniques to treat metastatic cancer [1,2]. Immunotherapy techniques have been tried to treat the cancers and improve other techniques with little or no toxicity to normal tissue. It has offered more safety to traditional chemotherapy with more long-term protection through immunological memory [1].

With knowledge of the relationship between immune cells, cancer cells, different tumor escape mechanisms and their functions have grown rapidly; a new technique based on this function of immune cells has been defined recently, which is known as adoptive cell transfer (ACT) [1–3]. This technique is the convergence of the knowledge from molecular biology and cellular immunology that in past decades has provided a new way for the treatment of cancer. Lymphocytes have played an important role in immune cancer therapy, hence, ACT for cancer therapy relies on the ex vivo generation of tumor-specific lymphocytes with high activity and their injection in large numbers to the host. In a half-century, different preclinical models have shown a successful ACT therapy not only for cancer therapy, in particular, melanoma but also for other disorders such as GVHD [4]. The scientists have also shown how the characteristic of lymphocytes is important for ACT. T cells, in particular cytotoxic T cell (CD8 + T cells) are important in adaptive immunity that has a straight effect on cancer cells and pathogen clearance [1].

Different reports have also indicated that how host immune environment is important to increase the ACT efficiency. Researchers have manipulated it using immunosuppression and IL-2 with transferred T cells before any cell administration [4].

There are three forms of ACT which has been developed for cancer therapy: (1) tumor-infiltrating lymphocytes (TILs), (2) T cell receptor (TCR) T cells, and (3) CAR T cells which are explained below.

5.1.1 Cancer therapy with TIL

TIL is a form of cellular therapy for cancer ACT [5,2]. This technique is the older form of ACTs using melanoma T cells. Melanoma was the first model that tumor-reactive T cells could expand in vitro to a large number for ACT. Steven Rosenberg was the first one who exposing tumor-derived lymphocytes to high dose IL-2 as an effective cancer treatment for patients with refractory metastatic melanoma [6,7]. To increase the effect of TIL, more studies have been done that in one of the patients received cyclophosphamide in one dose before cell administration and the result has shown that there was only 34% difference between patients given cyclophosphamide and not given that drug. Overall, all studies have indicated that the duration of responses is not more than one year and the survival after administration is very brief [8]. Other studies such as the use of fludarabine with cyclophosphamide have been done to change the method to increase the effect of the TIL, although in some aspects they were successful, but with all changes ultimately the objective response rate reached to 54%. However, the investigation has shown that with all the amendments this technique still has some limitations including the preparation of a large number of lymphocyte, the requirement for acute clinical management of toxicities, patient selection bias, and infrastructure demand [9]. Nevertheless, yet in patients with chemotherapy-refractory, TIL has a positive effect. Recently, different reports have indicated that chemotherapy before the administration can increase the response to adoptive immunotherapy with TIL [2]. More new techniques, such as in vitro activation of costimulatory ligands or PD1+ with tumor-reactive TIL, have led to getting a good result in a preclinical model of breast-ovarian and colorectal cancer [2].

5.1.2   Cancer therapy with TCR

Until 1990 ACT was limited to the infusion of lymphocytes to treat cancer and tumor cells. In 1990 by the innovation of gene engineering a new sight opens to the development of this technique [2].

On T cell surface there is an α- and a β-chain which noncovalently associated with the CD3 as a TCR. Activation of the receptor has happened once the TCR detects peptides bound to MHC of the surface of antigen-presenting tumor cells. And by detection, the T cells can destroy them and inhibit the tumor cell expression [4]. The first TCR cancer immunotherapy has been tested against metastatic melanoma. In this treatment scientists have used a TCR that bound to the human lymphocyte antigen A2 (HLA-A2) from a melanocytic antigen that makes a higher active TCR, targeting MART-1 (melanoma antigen recognized by T cell-1). This approach even could detect the malignant cells with lower MART-1 expression [10]. Although the development of this therapy using TCR was worthy, it also has other side effects which were targeting normal melanocyte of skin, eye, cochlea, hence, some off-tumor toxicity has happened in 50% of the patient who got this treatment. In others TCR cancer-based therapies which were the targeting of antigen MAGE-A3, some fatal neurotoxicity and cardiotoxicity have been observed, nonetheless, the treatment with targeting antigen NY-ESO-1 for cancer of the testis, a clinical efficacy without any cardiotoxicity has been reported [9]. This experiment has increased the hope to make this technique as an efficient method. Engineered T cell and its entrance to the CTR cancer therapy have shown that the production of specific T cells for particular neo-antigens would be safer and efficient than the share antigen. However, these ideas need clinical trials and lots of research to be done [2,4,10].

5.1.3   CAR T cell therapy

By the development of Gene-transfer techniques chimeric antigen receptor (CAR) T cells as immune cancer therapy has been developed [10,11]. The advantage of CAR is it combines antigen-binding domains and it has additional costimulatory domains for receptors such as CD28, OX40, and CD137 [10,4]. The problem of other techniques as they could not able to target tumor cells which represent less or does not represent MHC. It means that it has overcome some limitations of TCR T cell therapy which one of them was the need for MHC expression for tumor cells [12]. Kuwana and Eshhar [13,14] were the first group who have shown that these synthetic receptors molecules are independent of MHC for targeting. This property has been known as one of the best advantages of CAR T cells. Therefore, CAR T cells can kill many tumors cells and promote immune surveillance to prevent tumor recur by adaptive immunity (Table 5.1).

CAR T cells technique provides a broad development in cancer immune therapy. At first it only has been used in the treatment of leukemia and lymphoma but nowadays, with the help of engineered T cells and genetic editing treatment, it has gone far away from oncology and it has also entered the application of organ transplantation and treatment of autoimmunity [12].

Table 5.1 Characteristic of CAR T cells and TCR T cells [10].

CAR T cells	TCR cells
Signal amplification from synthetic biology: 200 targets can trigger CAR T cells	Sensitive signal amplification derived by evolution of the TCR
Avidity-controllable	Low-avidity, unless engineered
CAR T cells targets surface structures: proteins, glycan	TCR cells targets intracellular proteome
MHC-independent recognition of tumor targets	Requires MHC class I expression and HLA matching on tumor
Decade-long persistence	Lifelong persistence
Serial killers of tumor cells	Serial killers of tumor cells
Cytokine release syndrome more severe than TCR-based therapy	Off-tumor toxicity difficult to predict

Due to the high costimulatory of CAR T cells, it has enabled the persistent cells in cell therapy treatment to be one of the aims of CAR T cell therapy. The result of treatment with CAR T cells was disappointing until 2011 as the peripheral-blood T cell engineered with a CD19 was prepared and infused as a specific CAR T cell to destroy the lymphomas and leukemia in mice. Its positive result was promising for treatment using CAR T cell therapy [2]. The success of using CD19 CAR T cells has been observed in other leukemia and hematological malignancies. In 2018, Park et al. studied 53 patients who got anti-CD19-CAR T cells as a treatment for more than one year [15]. The result has indicated that 83% of the patient got complete remission in the median 29 months of follow up. Recently, a powerful antitumor activity using CAR T cell targeting CD22 in all hematological malignances and BCMA in myeloma have been reported. But all those have restriction to the B cell lineage that it causes toxicity because of the normal cell targeting. And until now targeting only tumor-associated antigens, remain limited. However, the combination of chemotherapy with CAR T cells could be promising, in particular for the patient with chemotherapy-refractory [15]. In 2015 Koshenderfer et al. have shown the effect of the treating chemotherapy-refractory-cell malignancies with anti-CD19 CAR T cells [16]. In this clinical trial research, 15 patients with advanced B-cell malignancies were investigated. They have received a conditioning chemotherapy regimen of cyclophosamide and fludarbine and single infusion of anti-CD 19 CAR T cells. The result has shown that eight patients got complete remission and four of them got partial remission. Also acute toxicity has been observed (fever, hypotension, delirium) but the result was anticipant. But it still needs further research and investigation in order to be set as a treatment method [16].

In general, one limitation of ACT is delivering targeted lymphocyte to tumor sites and sometimes its problem in expansion in the immunosuppressive tumor microenvironment [16]. In order to reduce and solve this problem using biomedical engineering, scientists have prepared a scaffold with alginate to implant it close or at resection sites [17].

They work on the mouse breast cancer resection model and showed how this scaffold effectively worked and how reduced tumor relapse compared to conventional delivery methods. They have also demonstrated how biopolymer scaffolds could evoke an antitumor immune response with the ability to eliminating widespread tumor metastases [17].

5.1.4 Dendritic cell-based therapy

Dendritic cells (DCs) are the types of antigen-presenting cells in the immune system [18,2]. Their primary function is the processing of antigens and the delivery of antigens in the primary response to naïve cells and the secondary response to memory B cells. There are mature and immature DCs. The main function of immature DCs is antigen-capturing, whereas for the mature DCs is antigen-presenting. In in vivo, what make the DCs mature is their ability to migrate from the peripheral tissue to draining lymphoid organ [18].

These DCs contain antigens that move to naïve T cells. It induces a cellular immune response which contains both CD4+ T cells and CD8 + T cells. Activation of Naïve and memory B cells by DCs makes them an important cell in transferring humoral immunity. They also can activate natural killer (NK) cell and natural killer T (NKT) cells [19].

The immunogenicity DCs have been shown in patients with cancer or chronic HIV infection.

DCs have a crucial role in determining the type of response that is induced. Activation of adaptive and innate immunity by DCs is an important ability that have been caused them to be used in vaccination in particular in cancer therapy [19,20]. Vaccination using DCs are able to induce an immunologic response, to increase the number of tumor antigens. Failure to delivery of tumor antigen to the target place is a limitation of cancer therapy. However, to overcome the limitation, DC-based therapy has been used in combination with other techniques.

By the generation of functional antigen-specific T cells researchers have also tried to solve this problem.

Despite some challenges regarding measuring the immunological effect of DC vaccination, some clinical trials have shown positive results but its cost is a crucial problem that still needs to be solved in order to this technique become useful for all patients [20].

Needs to treat cancer increase the attention of researchers to the innovation of new techniques with low cost to eradicate this mysterious disease. Adoptive cell therapy (ACT) recently has recognized as a method for cancer therapy that includes the administration of the immune cells with anticancer activity. Despite promising results that it has especially in CAR T cell therapy for the treatment of patients with melanoma, it has some limitations that in all three techniques are different. But in all, toxicity is the main problem due to the targeting of normal cells by these modified cells. Biomedical engineering, genetic engineering, and T cell engineering are needed to reduce this toxicity and increase its efficiency.

DCs also because of the capacity to activate adaptive and innate immunity has been considered in cancer therapy as well. However, there is limited knowledge about driving tumor antigens by DCs that first of all, there is a need to fill these gaps by lots of investigation and clinical trials to make DCs-based therapy as a high-efficiency technique for cancer therapy.

5.2   Innate cell-based therapy

5.2.1   Innate cells

Alongside the adaptive immune system described in the previous section, there is an innate immune system. The innate immune system consists of cells derived from both myeloid and lymphoid progenitors. Macrophages, DCs, neutrophils, eosinophils, basophils, and mast cells are among the innate immune cells which have been derived from the myeloid lineage. Alongside these cells mentioned above, there are another type of innate immune system cells derived from the lymphoid lineage known as innate lymphoid cells (ILCs); these cells are more prominent than the myeloid-derived cells in this section and include group 1 ILCs (ILC1s), group 2 ILCs (ILC2s), group 3 ILCs (ILC3s), NK cells, and lymphoid tissue-inducer (LTi) cells. This classification is schematically illustrated in Fig. 5.1.

The innate immune system and the use of innate cells to fight infections and tumors, despite the fact that these cells do not have antigen-specific cell surface receptors, have received considerable attention due to their variety and wide distribution [21].

In this chapter, we have focused more on ILCs because some of these cells have similarities to adaptive immune system cells and also have some antitumor properties.

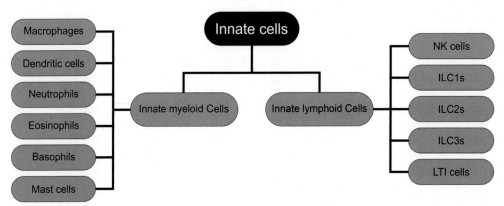

FIGURE 5.1 Classification of innate immune system cells derived from myeloid and lymphoid progenitors.

Table 5.2   Some of the cytokines and transcription factors expressed by ILCs.

Innate lymphoid cells	Secreted cytokines	Expressed transcription factors
NK cells	IFNγ, TNF	T-bet, EOMES
ILC1s	IFNγ, TNF	T-bet
ILC2s	IL-5, IL-9 and IL-13	GATA3
ILC3s	IL-22 and/or IL-17A	RoRγt

5.2.2   Innate lymphoid cells

As mentioned in the preceding section, ILCs are cells of the innate immune system that derived from lymphoid progenitors in the bone marrow. These cells are divided into five groups based on the cytokines they secreted and also on the expression of their transcription factors. For instance, IFNγ is a cytokine secreted by NK cells and also T-bet that expressed by this type of cells. Some of the secreted cytokines and transcription factors expressed by ILCs are summarized in Table 5.2.

In recent years, researchers have focused on ILCs due to the cell surface molecules expressed or secreted by these cells and their ability to counteract tumors. In this regard, studies have shown that NK cells induce immune responses through the release of various cytokines and chemokines and also have cytotoxicity to tumor cells. Although in the context of cancer treatment, more attention has been paid to NK cells, non-NK cell ILCs can also play a role in the fight against tumors due to the tumor microenvironment and the components of cytokines secreted in the tumor environment [22]. For example, in 2014, Mirchandani and colleagues have examined the similarities and effect of ILC2s and T cells on each other; the results of this study have shown that ILC2s by expressing MHC class II, introduce antigens to the adaptive immune system and induce T cell responses against antigens [23]. Or in another study conducted in 2018, the results have shown that in an animal model of mice lacked ILC2s genetically, the tumor growth rate and metastasis was increased dramatically; and the authors attributed this increase in tumor growth rate in the absence of ILC2s to IL-13 secretion by ILC2s, that induces cytotoxic T cell responses [24]. Although based on the researches demonstrating the effect of non-NK cell ILCs on stimulation of the adaptive immune system against antigens, their precise effect on cancer resistance is still unclear. And the main focus in this area is on NK cells.

5.2.3   NK cells

NK cells make up about 15% of blood lymphocytes. Morphologically they are similar to B and T lymphocytes but NK cells are larger and more granular. And most importantly, unlike T cells and B cells, they lack diverse antigen-specific receptors (TCR and BCR). The granules of these cells contain cytolytic proteins such as perforin and granulysin which give them cytotoxicity. In addition, like other ILCs, NK cells are capable of producing various chemokines and cytokines such as IFNγ and suppress tumors through various effective mechanisms such as the IFNγ-mediated pathway. Therefore, NK cells are able to distinguish normal cells from virus-infected cells and malignant cells and are able to kill them after attaching to these abnormal cells [25].

The function of these cells is due to the very complex balance between the inhibitory and activating signals that are transmitted into the cell by the inhibitory and activating receptors. When NK cells are exposed to normal cells, class I major histocompatibility complex (MHC) molecules present at the surface of normal cells bind to the inhibitory receptor of NK cells and NK cells are not activated and do not lyse normal cells. In contrast, when NK cells encounter virus-infected cells and malignant cells, the activating ligands present on the surface of these abnormal cells bind to the activating receptors of the NK cells and activate the NK cells and thereby lyse the virus-infected cells and malignant cells [26]. The function of these activating and inhibitory receptors in exposure to normal host cells and malignant cells is simply illustrated in Fig. 5.2. But in fact, the function of these receptors is much complex, because in addition to having

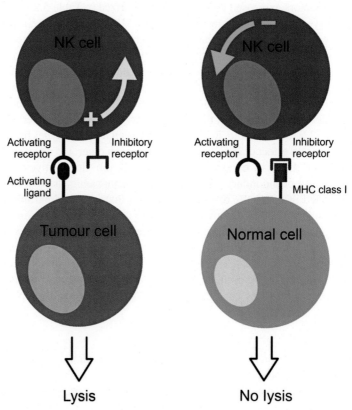

FIGURE 5.2 The function of inhibitory and activating receptors in NK cells in exposure to normal and tumoral cells.

a large number of activating an inhibitory receptor, and whether or not NK cells are activated depends on the response of all these receptors, some studies show that some tumoral cells alter their expression of MHC class I molecules.

NK cells in cancer immunotherapy have received much attention because they are widely cytotoxic and require a short time to lyse tumor cells and most importantly, NK cells that enter the patient's body in NK cell therapy do not need to strict HLA matching and also in NK cell therapy there is no possibility of graft-versus-host disease. And numerous clinical studies have shown that the use of allogeneic NK cells against various types of hematologic malignancies or solid tumors can be safe and effective.

Due to the advantages of using CAR-T cells and the success of these types of cells against neoplasms, which have been discussed in this chapter, researchers have moved toward the production of CAR-NK cells in order to increase the ability to kill the tumors in them. CAR-NK cell therapy has two major advantages over CAR-T cell therapy. First, the CAR-NK cells through their receptors can identify the target tumor cells, and unlike the CAR-T cell therapy that the tumor cells can escape through downregulating the CAR target antigen, in NK cell therapy, the tumor cells can rarely escape. Second, on the other hand, NK cells that enter the patient's body in NK cell therapy do not face immune rejection by the host [27].

5.2.4   Innate or adaptive immunity?

Recent researches have shown that NK cells have both innate and adaptive immune system properties. NK cells in the face of the tumors either kill them independently or cause T cells to be active and also in some cases, NK cells use both of these methods to eliminate tumor cells. According to investigations done to evaluate the efficacy of NK cell therapy as compared to the use of adaptive immune system cells against cancer, the results have indicated that although it has been reported in some cases that the process of purifying and uniformly dispersing of NK cells has been difficult, but in most cases and especially in recent years due to the same effectiveness of NK cells and the adaptive immune system cells, NK cell therapy has been preferred to other similar methods [28]. For example, in a study conducted in 2018, researches have compared the antitumor effects of iPSC-NK cells and CAR-T cells in an animal model of mice. The results of this study have shown that although the antitumor effects of both methods were almost identical, the mice that treated with NK-CAR-iPSC-NK cells had significantly longer survival and did not experience weight loss and other side effects such as increased levels of cytokines, unlike CAR-T cell therapy [29]. Other studies in recent years, in particular in 2018 and 2019, have indicated that researches in the field of cancer immunotherapy have switched to NK cell therapy because of the availability and lower cost of NK cell therapy than T cell therapy.

5.2.5   Preclinical and clinical examination of innate cell therapy

As noted in previous sections, despite all advantages of methods such as vaccination and targeting tumor cells through antibody-antigen binding which have been occurred by adaptive immune system cells, in recent years, preclinical and clinical examinations in the field of innate cell-based therapy has expanded dramatically. For instance, Xiaowen Tang et al. in a clinical study in 2018 have examined the safety of CD33-CAR NK cells in patients with acute myeloid leukemia (AML). In this phase I trial study, which has included three patients with advanced AML, each patient has received 5×10^9 CAR NK-92 cells. The results of this study have shown that no harmful effects were observed and the cost of using these cells was much lower than the use of CAR-T cells [30].

In another clinical study conducted in 2016, the researchers have investigated the effects of Healthy Donor-Derived Allogeneic NK cells in 17 patients with malignant lymphoma or with advanced solid tumors. In this study, each patient has received a maximum of 3×10^7 highly activated NK cells (MG4101) and the results have shown that this treatment was not only safe and effective but also effective in maintaining the effector arm of patient's immune system response against cancer [31].

In another study in 2019, the safety and efficacy of allogeneic NK cells injection in patients with hepatocellular carcinoma have been investigated. In this clinical experiment, 21 patients with primary hepatocellular carcinoma were enrolled. Allogeneic NK cells were donated by relatives of each patient and the follow-up time for patients was one year after NK cell therapy. The authors have reported that after NK cells infusion, 10 patients had a fever, one had a partial headache, and one patient had partial hepatalgia. And no other harmful event like graft-versus-host disease (GVHD) happened. And the overall results of this study have indicated that allogeneic NK cells are safe and effective in the treatment of primary hepatocellular carcinoma and also the antitumor properties of NK cells play an important role in activating the host's natural immune system and preventing the progression of the disease [32].

And in a preclinical study performed in 2017, about 5×10^6 human NK cells (NK-92MI) were injected twice into an animal model of mice that had pulmonary metastasis of anaplastic thyroid cancer (ATC). The results of these experiments have shown that NK cell therapy can be effective in targeting and treating anaplastic thyroid cancer (ATC) and prevent metastasis to other sites [33].

5.3   Progenitor stem cell therapy

Stem cells are kind of cells that have the ability of self-renewal and to the production of mature cells through differentiation [34]. Stem cells basically have categorized as 'embryonic' (ESCs) or 'somatic' (SSCs). Somatic stem cells that also known as adult stem cells are multipotent and can differentiate to any cell type. These are including neural stem cells, mesenchymal stem cells, hematopoietic stem cells, and endothelial progenitor cells. Investigation of functions and features of stem cells have opened a new sight into the therapeutic methods for different diseases in particular in cancer therapy [34,35]. Stem cells have the ability to migrate to solid tumors and micrometastatic lesions. This feature makes the stem cells as a specific antitumor agent.

Engineered stem cells can express a variety of antitumor agents and be a good help for conventional chemotherapeutic agents [36]. But there is some limitation on it particularly about the relationship between stem cells and normal cells which needs more investigation on fundamental stem cell mechanisms which is demonstrating the regulation of cancer cell proliferation because what makes cancer as a disease is the unregulated self-renewal of these cells. Knowledge about fundamental of the stem cells leads to improve therapeutics techniques and also it will improve stem cell-based regenerative medicine and anticancer techniques [36].

Among stem cells, there is a kind of cell that is known as progenitor stem cells. These cells today are shown to be useful in cancer immunotherapy [37]. Progenitor cells often confused with adult stem cells; progenitor cells like stem cells can differentiate to one or more kinds of cells but it cannot divide and reproduce unlimitedly. It can be said that they are between stem cells and fully differentiated cells. Their potency depends on the type of their parent stem cells and their niche. These cells like adult stem cells have the tendency to migrate toward the tissues where they are needed. They found in adult organs and their main task is to repair the body. Progenitor cells have little activity and they exhibit slow growth and mostly in the case of tissue injury, or dead cells they can be activated. What makes the progenitor cells to move and be activated is the growth factors and cytokines. Investigation of cytokines and growth factors of progenitor cells is important to use these cells as an antigen agent in cancer therapy [37,38].

Their limitation in differentiation and self-renewal make these cells to be more attractive in cancer therapy [34,37]. Lymphocyte progenitor cells and hematopoietic progenitor stem cells are important progenitor cells which are used in cancer therapy.

5.3.1 Hematopoietic progenitor cells in cancer therapy

Tumor microenvironment recently being the center of researches and it includes a component that it is shown they have an important role in the protection of cancer stem cells. Microenvironment signals also can effect on renewal and differential of normal cells. Hence, it is needed to understand the mechanism of all components involved in cancer cell environment to monitor the metastatic process [39].

Hematopoietic stem cells are progenitor cells that are found in blood and bone marrow. These cells act as a father cell and can be differentiated to the lymphocyte and myeloid cells [40]. These cells are contributed to the monitoring of the metastatic process. These cells have used in regenerative medicine and immunotherapy in cancer therapy. Recent research on the production and mobilization of these progenitor cells in patients with cancer and murine models have shown that hematopoietic progenitor can produce a strengthen immunosuppressive milieu around the premetastatic niche of distant regions [40]. This study was the first to track the development of hematopoietic progenitor cells and its differentiation to myeloid cells that has the suppressor effect on cancer cells. They present hematopoietic progenitor cells as a direct powerful tool in cancer therapy.

Self-renewal and differentiation abilities that progenitor cells have cause to use them in repairing the tissue after chemotherapy. Clinically it has been shown that transplantation of hematopoietic stem cells, will ease the lifelong hematological recovery after treatment with high dose radiotherapy or chemotherapy. These cells are collected from a donor and then injected to the patients with the goal to reconstitute bone marrow and to treat blood cells [39,40]. The collection of hematopoietic progenitor cells is different in some patients such as $CD34^+$ cells they collect with T cells and in others they have been collected from an allogenic donor. And each of them has a different method of collection [40].

The usage of progenitor cells in immune cancer therapy with the investigations of the TCR and CAR has been increased [41]. There are three reasons why progenitor cells are attractive to be used in TCR and CARs. First of all, HSCs have the ability of long-term blood cell production hence, HSCs can produce effective T cells engineered against the antigen of interest. They supply of T-lymphocyte progenitor cells that can increase the potential for the development of immunological memory. These cells are in contrast to the mature T cells that massively expanded before infusion. Second, one problem of TCR mechanism as it said before was the toxicity that it can provide due to targeting the normal cells. HSC can eliminate this target cytotoxicity because they are not arranged yet in their germ line. So it can provide a specific antigen for targeting. And the third one is the ability of HSC to mandate after high dosing of conditioning regimens to increase clinical efficacy. This feature can help to use HSC grafts as a way for Immunotherapy [41].

5.3.2 IPSC in cancer therapy

Induced pluripotent stem cells (iPSCs) are cells that have been derived from different sources such as skin or blood cells that have the ability to reprogram back to the first stage, for instance, embryonic progenitor stem cells [42]. This technique has widely been used in regenerative medicine. But still it has several problems that need to be solved. One of its problems is that it does not have the ability of robust differentiation protocol toward a specific tissue with the generation of the cells.

Another field where these cells are used is in cancer. The first study in cancer was on hematopoietic malignancy and then with the development of the novel model of myeloid proliferation. It has progressed rapidly and has been used in different cancer and more recently it is used in solid tumors and hereditary cancers [43].

As it was said earlier, the main goal of cancer immunotherapy is to stimulate a steady immune response against the tumor cells. The immunological checkpoint blockade is one of the new techniques that are used today as a new way in cancer immunotherapy [44]. The main purpose of them is to enhancing the tumor-specific activity of the immune system. To reach this goal, the immunoglobulins are the key points and cell receptors such as CTLA4 is working as activators or inhibitors to initiate an immune inhibitory signal in the targeted tumor cells. This technique is used as one of the cancer-immune methods but it has a problem in culturing and manipulating immune cells ex vivo hence what has recently been used are embryonic stem

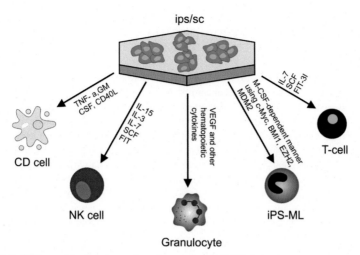

FIGURE 5.3 IPSCs and factors to produce immune cells [44].

cells due to the high ability they have to differentiate to immune system cells. IPSCs have the same properties of the EMCs, therefore, these cells are more suitable to be used in this immunotherapy technique. IPSC can be generated from somatic stem cells of the patients. These cells have the potential to generate a specific immune cell for the patients with antitumor activity that can be used in personal cancer treatment. Also, IPSCs does not have the problems related to ESC, for instance, undesired immune reaction against foreign tissue. Fig. 5.3 depicts some factors for generating immune cells from IPSCs.

With the development in adoptive T cell therapy for cancer, lack of available and antigen-specific human T lymphocyte becomes a hindrance for this technique [45]. Combination of IPSCc and CAR technologies offers a potential new source of T cells. This system could facilitate the production of T cell population with additional genetic modification and specification in cancer treatment. In 2013 Wakao et al., by the combination of IPSCs and CART cells, have generated human T cells targeted to CD19 in tissue culture. They have shown that IPSC inhibits tumor growth in a xenograft model [46].

In the following chapters, a brief review of bioengineering approaches in cancer treatments based on nonbiological methods such as laser, magnetic, and ultrasound will be discussed.

References

[1] V. Deschoolmeester, D. Kerr, P. Pauwels, J.B. Vermorken, Cell based therapy: modified cancer cells, Immunotherapy for Gastrointestinal Cancer, Springer, Champ, 2017, pp. 23–46.
[2] C. Yee, Adoptive T cell therapy: points to consider, Curr. Opin. Immunol. 51 (2018) 197–203.

[3] M.E. Dudley, S.A. Rosenberg, Adoptive-cell-transfer therapy for the treatment of patients with cancer, Nat. Rev. Cancer 3 (9) (2003) 666.

[4] M. Sadelain, I. Rivière, S. Riddell, Therapeutic T cell engineering, Nature 545 (7655) (2017) 423.

[5] S.A. Rosenberg, N.P. Restifo, J.C. Yang, R.A. Morgan, M.E. Dudley, Adoptive cell transfer: a clinical path to effective cancer immunotherapy, Nat. Rev. Cancer 8 (4) (2008) 299.

[6] S.A. Rosenberg, B.S. Packard, P.M. Aebersold, D. Solomon, S.L. Topalian, S.T. Toy, et al. Use of tumor-infiltrating lymphocytes and interleukin-2 in the immunotherapy of patients with metastatic melanoma. A preliminary report, New Engl. J. Med. 319 (1988) 1676–1680.

[7] S.A. Rosenberg, J.R. Yannelli, J.C. Yang, S.L. Topalian, D.J. Schwartzentruber, J.S. Weber, et al. Treatment of patients with metastatic melanoma with autologous tumor-infiltrating lymphocytes and interleukin 2, J. Natl. Cancer Inst. 86 (1994) 1159–1166.

[8] S.A. Rosenberg, P. Aebersold, K. Cornetta, A. Kasid, R.A. Morgan, R. Moen, et al. Gene transfer into humans—immunotherapy of patients with advance melanoma, using tumor-infiltrating lymphocytes modified by retroviral gene transduction, New Engl. J. Med. 323 (1990) 570–578.

[9] C.H. June, Adoptive T cell therapy for cancer in the clinic, J. Clin. Investig. 117 (6) (2007) 1466–1476.

[10] C.H. June, R.S. O'Connor, O.U. Kawalekar, S. Ghassemi, M.C. Milone, CAR T cell immunotherapy for human cancer, Science 359 (6382) (2018) 1361–1365.

[11] X. Wang, I. Rivière, Clinical manufacturing of CAR T cells: foundation of a promising therapy, Mol. Ther. Oncolytics 3 (2016) 16015.

[12] C.H. June, M. Sadelain, Chimeric antigen receptor therapy, New Engl. J. Med. 379 (1) (2018) 64–73.

[13] Y. Kuwana, Y. Asakura, N. Utsunomiya, M. Nakanishi, Y. Arata, S. Itoh, et al. Expression of chimeric receptor composed of immunoglobulin-derived V regions and T-cell receptor-derived Cregions, Biochem. Biophys. Res. Commun. 149 (1987) 960–968.

[14] Z. Eshhar, T. Waks, G. Gross, D.G. Schindler, Specific activation and targeting of cytotoxic lymphocytes through chimeric single chains consisting of antibody-binding domains and the gamma or zeta subunits of the immunoglobulin and T-cell receptors, Proc. Natl. Acad. Sci. USA 90 (1993) 720–724.

[15] J.H. Park, I. Rivière, M. Gonen, X. Wang, B. Sénéchal, K.J. Curran, et al. Long-term follow-up of CD19 CAR therapy in acute lymphoblastic leukemia, New Engl. J. Med. 378 (5) (2018) 449–459.

[16] J.N. Kochenderfer, M.E. Dudley, S.H. Kassim, R.P. Somerville, R.O. Carpenter, M. Stetler-Stevenson, et al. Chemotherapy-refractory diffuse large B-cell lymphoma and indolent B-cell malignancies can be effectively treated with autologous T cells expressing an anti-CD19 chimeric antigen receptor, J. Clin. Oncol. 33 (6) (2015) 540.

[17] S.B. Stephan, A.M. Taber, I. Jileaeva, E.P. Pegues, C.L. Sentman, M.T. Stephan, Biopolymer implants enhance the efficacy of adoptive T-cell therapy, Nat. Biotechnol. 33 (1) (2015) 97.

[18] H. Hasegawa, T. Matsumoto, Mechanisms of tolerance induction by dendritic cells in vivo, Front. Immunol. 9 (2018) 350.

[19] A. Gardner, B. Ruffell, Dendritic cells and cancer immunity, Trends Immunol. 37 (12) (2016) 855–865.

[20] W.W. Van Willigen, M. Bloemendal, W.R. Gerritsen, G. Schreibelt, I.J.M. de Vries, K.F. Bol, Dendritic cell cancer therapy: vaccinating the right patient at the righttime, Front. Immunol. 9 (2018) 2265, doi: 10.3389/fimmu.2018.02265.

[21] A.K. Abbas, A.H. Lichtman, S. Pillai, Cellular and Molecular Immunology E-Book, Elsevier Health Sciences, (2014).

[22] L. Chiossone, P.Y. Dumas, M. Vienne, E. Vivier, Natural killer cells and other innate lymphoid cells in cancer, Nat. Rev. Immunol. 18 (11) (2018) 671–688.

[23] A.S. Mirchandani, A.G. Besnard, E. Yip, C. Scott, C.C. Bain, V. Cerovic, et al. Type 2 innate lymphoid cells drive CD4+ Th2 cell responses, J. Immunol. 192 (5) (2014) 2442–2448.

[24] I. Saranchova, J. Han, R. Zaman, H. Arora, H. Huang, F. Fenninger, et al. Type 2 innate lymphocytes actuate immunity against tumours and limit cancer metastasis, Sci. Rep. 8 (1) (2018) 2924.

[25] R. Herberman, Natural cell-mediated immunity against tumors, Elsevier, (2012).

[26] E. Tomasello, M. Blery, E. Vely, E. Vivier, Signaling pathways engaged by NK cell receptors: double concerto for activating receptors, inhibitory receptors and NK cells, Paper presented at the seminars in immunology, (2000).

[27] H. Jiang, W. Zhang, P. Shang, H. Zhang, W. Fu, F. Ye, et al. Transfection of chimeric anti-CD138 gene enhances natural killer cell activation and killing of multiple myeloma cells, Mol. Oncol. 8 (2) (2014) 297–310.

[28] N. Tarek, D.A. Lee, Natural killer cells for osteosarcoma, Current advances in osteosarcoma, Springer, 2014, pp. 341–353.

[29] Z. Wang, Z. Wang, B. Li, S. Wang, T. Chen, Z. Ye, Innate immune cells: a potential and promising cell population for treating osteosarcoma, Front. Immunol. 10 (2019) 1114.

[30] Y. Li, D.L. Hermanson, B.S. Moriarity, D.S. Kaufman, Human iPSC-derived natural killer cells engineered with chimeric antigen receptors enhance anti-tumor activity, Cell Stem Cell 23 (2) (2018) 181–192.

[31] X. Tang, L. Yang, Z. Li, A.P. Nalin, H. Dai, T. Xu, et al. First-in-man clinical trial of CAR NK-92 cells: safety test of CD33-CAR NK-92 cells in patients with relapsed and refractory acute myeloid leukemia, Am. J. Cancer Res. 8 (6) (2018) 1083.

[32] Y. Yang, O. Lim, T.M. Kim, Y.O. Ahn, H. Choi, H. Chung, et al. Phase I study of random healthy donor–derived allogeneic natural killer cell therapy in patients with malignant lymphoma or advanced solid tumors, Cancer Immunol. Res. 4 (3) (2016) 215–224.

[33] Y.B. Xie, J.Y. Zhang, M. DU, F.P. Meng, J.L. Fu, L.M. Liu, et al. Efficacy and peripheral immunity analysis of allogeneic natural killer cells therapy in patients with hepatocellular carcinoma, Beijing Da Xue Xue Bao Yi Xue Ban (Journal of Peking University. Health Sciences) 51 (3) (2019) 591–595.

[34] L. Zhu, X.J. Li, S. Kalimuthu, P. Gangadaran, H.W. Lee, J.M. Oh, et al. Natural killer cell (NK-92MI)-based therapy for pulmonary metastasis of anaplastic thyroid cancer in a nude mouse model, Front. Immunol. 8 (2017) 816.

[35] T. Reya, H. Clevers, Wnt signalling in stem cells and cancer, Nature 434 (7035) (2005) 843.

[36] C.L. Zhang, T. Huang, B.L. Wu, W.X. He, D. Liu, Stem cells in cancer therapy: opportunities and challenges, Oncotarget 8 (43) (2017) 75756.

[37] A. Borah, S. Raveendran, A. Rochani, T. Maekawa, D.S. Kumar, Targeting self-renewal pathways in cancer stem cells: clinical implications for cancer therapy, Oncogenesis 4 (11) (2015) e177.

[38] H. Zhu, Y.S. Lai, Y. Li, R.H. Blum, D.S. Kaufman, Concise review: human pluripotent stem cells to produce cell-based cancer immunotherapy, Stem Cells 36 (2) (2018) 134–145.

[39] S.S. Khan, M.A. Solomon, J.P. McCoy Jr., Detection of circulating endothelial cells and endothelial progenitor cells by flow cytometry, Cytometry B Clin. Cytom. 64 (1) (2005) 1–8.

[40] A.J. Giles, C.M. Reid, J.D. Evans, M. Murgai, Y. Vicioso, S.L. Highfill, et al. Activation of hematopoietic stem/progenitor cells promotes immunosuppression within the pre–metastatic niche, Cancer Res. 76 (6) (2016) 1335–1347.

[41] E. Gschweng, S. De Oliveira, D.B. Kohn, Hematopoietic stem cells for cancer immunotherapy, Immunol. Rev. 257 (1) (2014) 237–249.

[42] P. Sachamitr, S. Hackett, P.J. Fairchild, Induced pluripotent stem cells: challenges and opportunities for cancer immunotherapy, Front. Immunol. 5 (2014) 176.

[43] Z. Jiang, Y. Han, X. Cao, Induced pluripotent stem cell (iPSCs) and their application in immunotherapy, Cell. Mol. Immunol. 11 (1) (2014) 17.

[44] F. Rami, H. Mollainezhad, M. Salehi, Induced pluripotent stem cell as a new source for cancer immunotherapy, Genet. Res. Int. (2016) 2016.

[45] Patel M. and S. 6, no. 3, pp. 367–380, 2010.

[46] H. Wakao, K. Yoshikiyo, U. Koshimizu, T. Furukawa, K. Enomoto, T. Matsunaga, et al. Expansion of functional human mucosal-associated invariant T cells via reprogramming to pluripotency and redifferentiation, Cell Stem Cell 12 (5) (2013) 546–558.

Laser-assisted cancer treatment

6

Chapter outline

6.1 Predicting the optical properties

There are several models to numerically predict the optical properties of various geometries such as Mie theory, boundary element, discrete dipole approximation, and finite element method. Mie theory and discrete dipole approximation will be discussed in the following.

6.1.1 Mie theory

This theory was introduced by Gustav Mie in 1908 [1]. This model is the analytical solution of Maxwell's equations to calculate the scattering properties of spherical and nonmagnetic particles. Maxwell equations are:

$$\nabla \times H(r,t) = \varepsilon_p(r) \frac{\partial E(r,t)}{\partial t}, \tag{6.1}$$

$$\nabla \times E(r,t) = -\mu(r) \frac{\partial H(r,t)}{\partial t}, \tag{6.2}$$

$$\nabla \cdot E(r,t) = 0, \tag{6.3}$$

$$\nabla \cdot H(r,t) = 0, \tag{6.4}$$

where H, E, ε_p, and μ are the magnetic field, the electric field, the electric permittivity, and the magnetic permeability, respectively. By applying Fourier transform to the Eq. (6.4), the time variable in electromagnetic field meets the Helmholtz equations [1]:

$$\nabla^2 E + k_1^2 E = 0, \tag{6.5}$$

$$\nabla^2 H + k_1^2 E = 0, \tag{6.6}$$

where wave vector is $k_1^2(r) = \frac{\omega(r)\varepsilon_p(r)\mu(r)}{c^2}$ while ω and c are the frequency and speed of light, respectively. By assuming the spherical harmonics shapes, E and H are replaced by vectors M_n and N_n. The subscript, n, represents different vector spherical harmonics to describe the polar contribution to the scattered fields ($n = 1$ dipolar, $n = 2$ quadripolar). The scattered field is calculated using series below [1]:

$$E_{sca} = \sum_{n=1}^{\infty} E_n \left(ia_n N_{ein} - b_n M_{oin} \right), \tag{6.7}$$

$$H_{sca} = \sum_{n=1}^{\infty} E_n \left(ib_n N_{oin} - a_n M_{ein} \right), \tag{6.8}$$

where $E_n = i^n E_0 \left(\frac{2n+1}{n(n+1)} \right)$ and E_0 is the incident filed. The subscript of o and e are odd and even parameter of the azimuthal solution to the vector. The a_n and b_n represent Mie coefficients. These values are calculated as fallow [1]:

$$a_n = \frac{m\psi_n(mx)\psi_n'(x) - \psi_n(x)\psi_n'(mx)}{m\psi_n(mx)\xi_n'(x) - \xi_n(x)\psi_n'(mx)} \tag{6.9}$$

$$b_n = \frac{\psi_n(mx)\psi_n'(x) - m\psi_n(x)\psi_n'(mx)}{\psi_n(mx)\xi_n'(x) - \xi_n(x)\psi_n'(mx)} \tag{6.10}$$

where ψ_n and ξ'_n are the Bessel function. M is the ratio between dielectric function of the nanoparticles and surrounded medium and X is the size perimeter: $x = \dfrac{2\pi n_m R}{y}$. Based on Drude-Lorentz model [2]:

$$\varepsilon = \varepsilon_B - \frac{\omega_p^2}{\omega^2 + \Gamma^2} + i\frac{\omega_p^2 \Gamma}{\omega(\omega^2 + \Gamma^2)},\tag{6.11}$$

where ε and ε_B are the dielectric of the nanoparticles and bulk material, respectively. $\Gamma = \Gamma_0 + \dfrac{vF}{R}$ is the damping term (R is the radius of the particle and vF is the Fermi velocity) and $\omega_p = \sqrt{\dfrac{4\pi n_e e^2}{\varepsilon_0 m_e}}$ is the plasma frequency of the free electron gas (m_e is the mass of the electron, n_e is free electron density, e is electron charge and ε_0 is the vacuum permittivity).

To calculate the amount of scattering, first, need to calculate absorption, extinction and scattering efficiencies. Based on Mie scattering coefficients:

$$Q_{ext} = \frac{2}{x^2} \sum_{n=1}^{\infty} (2n+1) R_e(a_n + b_n),\tag{6.12}$$

$$Q_{sca} = \frac{2}{x^2} \sum_{n=1}^{\infty} (2n+1)(a_n^2 + b_n^2),\tag{6.13}$$

$$Q_{abs} = Q_{ext} - Q_{sca}\tag{6.14}$$

ABSCAT and MATLAB program can assist to solve Eqs. (6.12)–(6.14) [2, 3].

6.1.2 Discrete dipole approximation

Discrete dipole approximation model was first introduced by Purcell and Pennypacker in 1973. This model provides the approximation for absorption, extinction, and scattering cross-section. In this method, each particle defines as a mass composed of finite number of elements with dipole interactions. Maxwell's equations are converted to a simple algebraic equation using Draine and Flatau methods [4]. For each dipole, the dipole moment of the external electric field and internal electric field causing by the neighboring dipoles are calculated. The electric field of one dipole is calculated as follow:

$$E_{dipole} = \frac{e^{\frac{iwd}{c}}}{4\pi\varepsilon_0}\left[\frac{\omega^2}{c^2 d}\hat{r} \times p \times \hat{r} + \left(\frac{1}{d^3} - \frac{i\omega}{cd^2} \right)[3(\hat{r} \cdot p)\hat{r} - p] \right],\tag{6.15}$$

where d and $\hat{r}$ are distance to the sampling point and unit vector of $\hat{r}$ which h is taken from dipole to the location at which electric field is collected, respectively. P is the vector related to the induce dipole moment,

$$P_i = \alpha_i E_{local},\tag{6.16}$$

where α_i is the polarizability of the materials and E_{local} is the electric field corresponding to the incident wave, $E_{inc,i}$, and the effect of the electric field of the other dipoles [4]:

$$E_{local} = E_{inc,\,i} + \sum_{\substack{j=1 \\ i \neq j}} E_j = E_{inc,\,i} + \sum_{\substack{j=1 \\ i \neq j}} A_{ij} P_j, \qquad (6.17)$$

where A_{ij} is the interaction matrix between all the dipoles in the particles. The scattering efficiency is:

$$Q_{sca} = \frac{k}{\pi r^2 |E_i|^2} \int_{dS} |E_{sca}(r)' \cdot \hat{r}|^2 dS. \qquad (6.18)$$

Fourier transform and DDSCAT are the applicable methods to solve Eq. (6.18).

6.2 Photothermal therapy

Photothermal therapy (PTT) refers to using photothermal agents to selectively destroy cancer cells by converting optical energy into heat. When photothermal agents expose to light, the electrons transfer from the ground state to the excited state. Then, the excited electron suddenly relaxes through nonradiative decay channels which leads to increasing the kinetic energy and subsequently raising the temperature around the targeted tissue. This heat is applied to destroy cancer cells. Several kinds of the photoabsorbing agent have been used in PTT including natural [5] or synthetic chromophores [6], noble metals [7], and carbon-based nanostructures [8]. In comparison to synthesize chromophores such as naphthalocyanines and indocyanine green, natural chromophores suffer from very low absorption cross-section. However, synthetic dye molecules exhibit photobleaching during laser irradiation. There has been a tremendous effort to develop nanoparticles for PTT.

The noble metals are considered a great deal of attention due to their four to five times higher cross-section absorption compare to the synthetic chromophores. Recently, gold nanoshells (GNSs), gold nanocages (GNCs), gold nanorods (GNRs), gold nanostars (GNSTs), carbon nanotubes, and graphene oxide (GO) have been considered as versatile nanoparticles for PTT on account of their strong absorption in near-infrared (NIR) region, high stability, and acceptable biocompatibility. Fig. 6.1 is shown these nanoparticles which is further described in the following section.

6.2.1 Gold nanoshells

GNSs compose of a dielectric core surrounded by a thin layer of gold. The surface plasmon resonance (SPR) of GNSs is related to the plasmon difference between the inner and outer shells. By changing the ratio of core radius to shell thickness, SPR wavelength is tuned from visible to NIR [9]. The typical method to fabricate GNSs is seed-mediated growth which composed of two steps—gold nanoparticle is first attached to the dielectric core and then growth under the specific condition to form a shell. The main studies of using GNSs for PTT are summarized in Table 6.1. GNSs, with the trade name of AuroShell is currently being examined for clinical study [10]. When AuroShell injects intravenously in a patient, will accumulate in tumors due

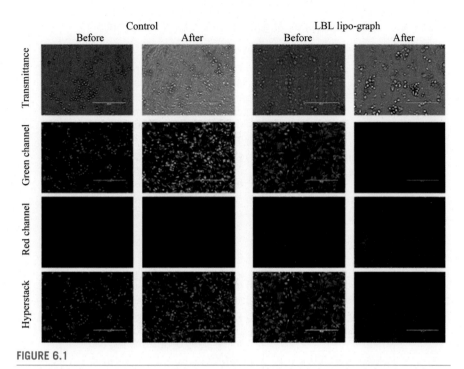

FIGURE 6.1

(A) The schematic of destroying the abnormal cells intravenously injected and accumulated in tumors via EPR effect. (B) Most commonly used nanoparticles for PTT.

to the enhanced permeability and retention effect (EPR). Then, the desired area is exposed to NIR irradiation. There is a claim that AuroShell can be able to selectively destroy cancer cells with minimum damage to the normal surrounded tissues.

6.2.2 Gold nanorods

Elongating to resemble rods, GNRs exhibit two plasmon resonance in transverse and longitudinal aspect. The transverse SPR mode is located at around 520 nm while the longitudinal SPR mode appears in a wide range of visible to NIR wavelength depending on the length-to-width aspect ratio. By increasing the aspect ratio, the SPR longitudinal peak shifted from blue to red. The optical properties of GNR is explained by Gans theory—an extended version of Mie theory with geometric parameters consideration [16]. Well-designed GNR with maximum absorption spectra in NIR region is absorbed great attention for PTT. Table 6.1 is summarized some of the significant studies in this field of interest. The main drawback of utilizing GNR is related to the presence of cytotoxic cetyltrimethylammonium bromide (CTAB) used in synthesis process. To tackle the problem, thiolated molecules have been introduced to replace CTAB. In 2009, Akiyama et al. showed that replacing thiolated molecules

Table 6.1 PPT application of gold nanostructures in NIR region.

Type	Targeting agent	Results	Ref
GNS	Biological: PEG Physical: MRI	(1) Colloidal stability and blood circulation time increased due to the presence of PEG molecules (2) Irreversible tissue damage after 4–6 min of laser irradiation (3) Tumor shrinkage	[11]
	Biological: anti-Her2 antibody conjugated PEG	(1) Dual imaging and therapy properties achieved by controlling scattering and absorption features of the GNSs (2) Selective destruction of carcinoma cells due to the presence of Her2 antibody	[12]
	Biological: anti-Her2 nanobody and PEG	(1) Selective tumor cell damage occurred in in vivo cell culture medium	[13]
	Cystine conjugated A54 peptides	(1) good targeting ability (2) low cytotoxicity to healthy cells	[14]
	Physical: superparamagnetic iron oxide Biological: C225 antibody, PEG	(1) Both superparamagnetic and optical properties in one platform which lead to the ability of dual imaging and therapy (2) Selective targeting to cancerous cells compared with nan-targeted nanoparticles in in vivo model (3) Selective PTT treatment	[15]
GNR	Anti-EGFR antibodies, poly(sodium-*p*-styrenesulfonate)	(1) Selectively targeted cell lines with overexpressed EGFRs (2) Malignance cells destroyed in half of the power densities needed to kill normal cells	[16]
	PEG	(1) Targeted photothermal destruction in in vivo model (2) Dramatic tumor volume reduction	[17]
	CD11b antibodies	(1) Efficient and selective cell death in in vitro model	[18]
	Anti-*Toxoplasma gondii* antibodies	(1) Efficiently targeted and kill parasitic protozoans	[19]
	PEG	(1) Administrating of X-ray scattering properties of GNRs to guide X-ray scattering photothermal agent to desire side (2) Increase the PPT efficiency	[20]
	Anti-EGFR antibodies, polydopamin	(1) Polydopamin improve the GNRs modification	[21]

Table 6.1 PPT application of gold nanostructures in NIR region. (*Cont.*)

Type	Targeting agent	Results	Ref
GNC	Anti-Her2 antibody, PEG	(1) Selective photothermal destruction of cancer cells	[25]
	Anti-Her2 antibody	(1) The number of GNSs were needed per cell were optimized (2) The power density of laser and exposure time were optimized to improve PTT outcomes	[26]
	PEG	(1) After one minute of laser irradiation, the temperature of the tumor increased to 50°C (2) Irreversible cell damage and metabolic change in mice model were observed after exposing NIR laser irradiation (3) Utilizing GNSs as a photoacoustic contrast agent	[27]
	Single wall carbon nanotube, A9RNA aptamers	(1) Selective photothermal destruction of prostate cancer cells was observed	[28]
GNST	Anti-Her2 antibody	(1) Successful targeting to cancer cells (2) Cancerous cells were destroyed after 5 min of 660 nm laser irradiation	[31]
GNST	PEG	(1) 5 min of continuous laser irradiation with power density of 15 W/cm^2 at 980 nm wavelength were sufficient to destroy abnormal cells lines in in vitro (2) After 2 days, intravenous injection of the nanoparticles led to their accumulation in desired side (3) 10 min of laser irradiation with power density of 1.1 W/cm^2 at 785 nm in animal model was able to effectively killed cancer cells	[32]
	PEG, transactivator of transcription (TAT)	(1) Effective penetrating to cancer cells in in vitro model (2) Effective cells ablation was observed after low energy laser irradiation	[33]

with CTAB in the conjugation process of PEG molecules improved circulation time and retention ability in the tumorogenic side [22]. Another strategy to improve the application of GNR in PTT is the construction of a silica shell around the GNR which not only increases the thermal stability of GNR during NIR laser exposure but creates the reservoirs for loading of hydrophobic drugs as well [23]. Also, the presence of silica on the surface improves further modification by employing silane coupling reactions. To increase the PTT efficiency, the large amounts of GNRs need to accumulate in the desired region. Several approaches have been studied to meet this goal, one of which was layer-by-layer assembly techniques composed of positive charged layer of poly(diallyldimethylammonium chloride) and negative charged layer of poly(sodium-*p*-styrenesulfonate). Qiu et al. claimed that the self-assembled GNR increased the nanoparticles accumulation and reduced the toxicity compared to nonassembled GNR [24].

6.2.3 Gold nanocage

GNCs with hollow interiors and porous walls were first introduced by Sun et al. The general method to synthesize GNCs is based on galvanic replacement reaction which occurred between nanocubes like silver template and gold precursor. The difference in electrochemical potential between these two leads to the depositing of the gold nanoparticles on the template. The SPR peak adjusts in the range of 600–1200 nm by controlling the amounts of precursors. Some studies focused on the application of GNS in PTT are reviewed in Table 6.1.

Not only GNSs have been used for PTT but also considered for dual PTT/PDT. Khlebtsov et al. functionalized the surface of silica-coated GNCs using a PDT sensitizer, Yb-2,4-dimethoxyhematoporphyrin [29]. Their introduced nanocomposite produced both free radical oxygen and heat during laser emission. Also, the presence of Yb-2,4-dimethoxyhematoporphyrin in the structure exhibited NIR luminescence originated from Yb^{3+}. In another study, hypocrellin B photosensitizer was incorporated into the lipid-coated GNCs [30]. Synergistic PTT and PDT effects were observed during pulse laser irradiation on HeLa cells.

6.2.4 Gold nanostars

GNSTs are novel stars shape nanoparticles composed of a small core and multiple sharp tips, the optical properties of which related to the size of the tips. The strong SPR effect of GNSTs is mainly related to the SPR difference between the core and the tips. The main method to produce GNSTs is typically seed-mediated growth method or seedless procedure using 2-[4-(2-hydroxyethyl)-1-piperazinyl] ethanesulfonic acid (HEPES) having a role of both reducing agent and shape controller. The main studies of using GNSTs in PTT summarized in Table 6.1.

6.2.5 Carbon nanotubes

Compared to metallic nanoparticles, carbon nanostructures have several advantages. First, the large number of functional groups which allow proper surface modification to increase stability and blood circulation, thereby increasing drug accumulation in the tumorogenic side. Second, strong optical absorption in NIR region with high penetration depth to soft tissues.

Among all the carbon nanostructures, single-walled carbon nanotubes (SWCNTs) were the first ones introduced for PTT application. Delivering SWCNTs to the desired region is facilitated by intravenous or intratumoral injection. Choi et al. investigated the effect of intravenously injection of PEG-conjugated SWCNTs on human epidermal mouth tumor in mouse [34]. Their results demonstrated that compared to nonirradiated laser group NIR irradiation significantly decreased tumor volume. To increase cellular uptake, Robinson et al. functionalized SWCNTs using PL-PEG and C18-PMH-PEG which were intravenously injected into mice with xenograft tumors. Their nanoparticle provided the opportunity of both imaging and therapy in a single platform [35]. Exploiting lower power density led to produced clear fluorescence image while administrating higher power density led to destroy tumor cells. The author claimed that by using their introduced nanoparticles, the power densities needed to kill abnormal cells successfully was much lower than the corresponding amount that was necessary by using nanorods. To improve PTT application of SWCNTs, carbon nanotube sorting technique was utilized to separate particular chirality with the best performance in NIR region by density gradient ultracentrifugation. Antaris et al. found that density gradient ultracentrifugation could separate carbon nanotubes with 80% purity [36]. After high-purity separation, SWCNTs surfaces were modified using PL-PEG polymers. Low dose SWNTs, 0.254 mg/kg, was sufficiently enough for tumor shrinkage during exposure of ultralow irradiation power density of 0.6 W/cm^2 which has to do with not only removing all SWCNTs with no absorbance in NIR region but improving cellular uptake of separated-SWCNTs by surface modification as well [36]. The authors claimed that around 4 μg of SWCNTs per mouse were enough for both imaging and therapy which was the lowest amount of PTT agent being reported for both therapy and diagnostic purpose in a mouse model [36].

6.2.6 Graphene

Graphene is a two-dimensional allotrope of carbons bonding together hexagonally. Having strong absorption cross-section in NIR window, graphene has been successfully utilized for PTT application. In 2010, Liu et al. studied the effect of PEGylated nanographene sheets injected intravenously as a PTT agent [37]. PEG was covalently attached to six-armed amine groups and then attached to the surface of nanographene oxide (GO). This modification not only improved water solubility of graphene but provided the opportunity for further surface modifications with fluorescent dyes as well. An account of high modified-graphene accumulation in the tumorogenic region in one hand and strong NIR absorption in the other hand, the introduced nanoparticles destroyed cancer cells in an efficient manner during laser irradiation. Histological analysis indicated no cellular toxicity [37]. Although the amounts of graphene needed for PTT have been higher than SWNTs, the cellular uptakes of SWCNTs have been lower than graphene motivating to consider graphene as a good candidate for PTT application. To decrease the amount of graphene needed for PTT therapy, several approaches have been accomplished. In 2010, Robinson et al. reduced graphene oxide to graphene to restore sp^2 carbon nanostructures leading to an increase

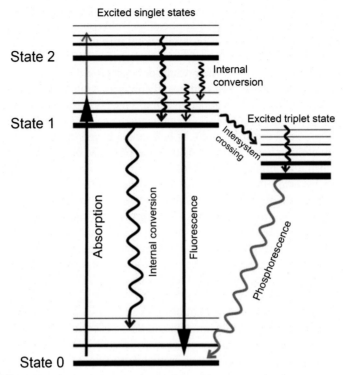

FIGURE 6.2

(A) Cells viability after incubation with GO (*red*) and rGO-Arg (*black*), (B) cell viabilities after NIR irradiation in the presence of 100 and 400 µg/mL of rGO-Arg and GO, respectively, (C) live/death assay in the presence of rGO-Arg. *Green* channel represented live cells, *red* channel represented dead cells.

in electrical conductivity [38]. In another study, Hashemi et al. conjugated Arginine to graphene oxide and then reduced it by adding hydrazine monohydrate (rGO-Arg). After reduction, the color of the suspension changed from brown to black, indicating an increase in the absorption from visible to NIR. They also obtained cross-section absorption value for rGO-Arg and compared to the value reported for graphene oxide. The cross-section absorption of rGO-Arg was 3.2 times higher than graphene oxide. Although no in vivo experiment was done on their introduced nanoparticles, the in vitro data suggested an efficient cellular death during NIR pulsed laser irradiation. Cell toxicity assays were represented in Fig. 6.2. As shown in Fig. 6.2A, rGO-Arg exhibited higher cell viability than GO which has to do with a large amount of negative charge on the GO surface being decreased during Arg conjugation. PTT potential of rGO-Args was examined using MTT-assays (Fig. 6.2B). The Toxicity of both GO and rGO-Args was increased by increasing exposure time from 2 to 4 min. Fig. 6.2B shown that rGO-Arg had a significant PTT effect compared to GO, even after 4×

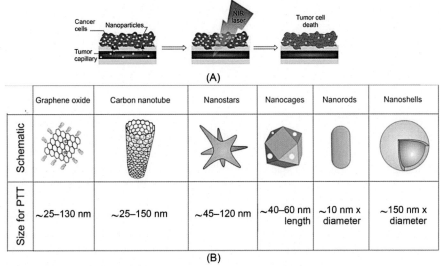

	Graphene oxide	Carbon nanotube	Nanostars	Nanocages	Nanorods	Nanoshells
Schematic						
Size for PTT	~25–130 nm	~25–150 nm	~45–120 nm	~40–60 nm length	~10 nm x diameter	~150 nm x diameter

(B)

FIGURE 6.3 When light absorbed by photosensitizers, the electrons move from low energy to high energy. The excited electron may lose the energy either by Internal conversion (nonradiative decay) or emitting fluorescence.

increasing in the concentration of GO. In addition, live-death assays provided supporting data on the potential application of rGO-Arg shown in Fig. 6.2C. The red spot represented cells dead increased by increasing NIR laser exposure time [39].

6.3 Photodynamic therapy

Photodynamic therapy (PDT) or photochemotherapy, is a kind of treatment involving photosensitizer and light to destroy abnormal cells during the formation of free radicals in photochemical and photobiological processes. When visible or NIR light exposes to photosensitizers, it reacted with oxygen of tissue and produced toxic singlet oxygen or fluorescent (Fig. 6.3). The first photosensitizer was acridine, reported in 1900 [40]. Shortly, eosin was introduced for skin cancer treatment [40]. Although a large variety of photosensitizer has been reported for PDT, porphyrin-based sensitizers has absorbed great attention especially in the clinical application which has to do with their high yields of free radical production and high retention time in tumorogenic region. Photofrin has been approved for clinical usage by FDA. Although porphyrin has considerable characteristic features for PDT, its application is limited to superficial organ since its maximum absorption is less than 640 nm in wavelength. To tackle the issue, the second generation of photosensitizers have been introduced, whose maximum absorption is in NIR regions such as chlorins, core-modified porphyrins, naphthalocyanine and phthalocyanine.

In addition to the tendency to be absorbed in NIR region and nontoxicity in the absence of NIR irradiation, photosensitizer must successfully be accumulated in the desired side and also it should be able to induce cell death via activating of dendritic and T cells. In the following section, some photosensitizers making the above criteria will be described.

6.3.1 Methylene blue

Methylene blue (MB) is one of the organic dyes being able to kill malignant cells. Owing to its positive charged and low molecular weight, the interaction of MB with mammalian cells increased lead to be chosen as a suitable candidate for PDT application. Although MB commonly applied in antimicrobial PDT, there have been several studies focusing on the application of MB in PDT due to its cost-effectivity and availability. Samy et al. utilized MB for PDT to study the outcomes on basal cell carcinoma in clinical stage. Complete treatment with good cosmetic outcomes was observed in 11 out of 17 patients [41]. Currently, the application of MB is limited due to the lack of stability after intravenous administration leading to the reduction of leuco MB with no photosensitizing properties [42]. Encapsulation of MB in nanoparticles to maintain their activity during blood circulation can be a solution.

6.3.2 Photogem

Photogem is originated from hematoporphyrin being one of the first generations of the photosensitizers with photochemical properties similar to Photofrin. Photogem got approved for clinical usage from "Brazilian Health Surveillance Agency (ANVISA)" and "Pharmacology State Committee of the Russian Federation". The maximum absorption of Photogem is between 500 and 630 nm. Its main side effect is related to the long-lasting photosensitivity in the body after several weeks of administration decreased by incorporating it into the drug delivery system. Low absorption spectra of the Photogem in the therapeutic window and long-lasting photosensitivity led to introduce the second generation of photosensitizes including chlorins and 5-aminolevulinic acid [43].

6.3.3 Chlorins

Chlorins with strong light absorption in the therapeutic window mainly between 640 and 700 nm have absorbed great attention due to its natural origin. Being hydrophilic reduced types of porphyrins, chlorins structure is similar to porphyrins. One of the most studies chlorin used as a photosensitizer is mono-L-aspartyl chlorin e6 which has two significant features including strong absorption between 650 and 680 nm and a high yield of free radical production. Another promising photosensitizer belonged to chlorin family is Photodithazine. It is one of the second generations of the photosensitizing agent with little skin photosensitivity [44]. Chlorins have been widely used for oral cancer treatment [45]. In 2010, Parihar et al. investigated the effect of PDT on carcinoma cells using chlorins photosensitizers in animal models. Their results demonstrated complete cell ablation in in vitro model and tumor shrinkage in in vivo model a week after PDT [45]. Despite having very good features in PDT therapy, chlorins need to be incorporated in a delivery system to improve its accumulation and blood circulation time such as chitosan, silica, and polymeric nanoparticles.

6.3.4 Curcumin

Curcuma longa L. dived polyphenolic compound, curcumin has been investigated for a variety of applications including cancer treatment, antimicrobial diseases, wound treatment, and joint inflammation. Curcumin with a wide absorption spectrum from 300 to 500 nm has a strong potential as a photosensitizer especially for superficial infections [46] and superficial tumors such as oral cavity and skin [46]. Owing to poor penetration depth, curcumin is not a good photosensitizer for deeper lesion. One of the main drawbacks of using curcumin in PDT is its poor solubility leading to poor bioavailability and low pharmacological outcomes. To tackle this challenge, nanotechnology has been applied to improve aqueous solubility. In 2012, Mohan Yallapu et al. encapsulated curcumin into cellulose nanoparticles and studied the effect of curcumin assisted PDT in prostate cancer treatment [47].

6.3.5 Phthalocyanines

Phthalocyanines are one of the second generations of photosensitizers consists of four isoindole groups attached together in a larger ring. It has unique properties including ease of synthesis and modification, high stability, long and strong absorption wavelength and high yield of free radical production. The physiochemical properties can be modified either by substitution of cationic metals in the core or by peripheral modifications to synthesis hybrid photosensitizer. Although phthalocyanines have great potentials for photosensitizer, they suffer from lack of tumor specify which can be tackled by conjugating to tumor-targeting peptides or incorporating into nanoparticles. Muehlmann et al. incorporated aluminum-phthalocyanine chloride into poly(methyl vinyl ether-co-maleic anhydride) nanoparticles leading to 10 times increase in PDT potential compared to its free form [48].

6.3.6 Hypericin

Hypericin is a red plant pigment being applied in traditional medicine for centuries. There is an evidence indicating hypericin has a potential to treat cancer, depression, and viral. Hypericin can be considered for PDT due to several excellent properties, including light absorption spectrum near to NIR, high yield of free radical production and tumor selectively, low toxicity, high stability. Several studies examined the application of hypericin for PDT. Kleeman et al. studied the application of hypericin assisted PDT for destroying melanoma cells [49]. In another study, Barathan et al. examined the effect of hypericin to kill hepatocellular carcinoma cells

Table 6.2 Photosensitizer agent under clinical or clinical approved.

Photosensitizers	Type of cancers	Ref.
Porfimer sodium	Lung cancers	[52, 53]
	Esophageal cancer	
	Bladder cancer	
	Cervical cancer	
	Skin diseases	
	Kaposi's sarcoma	
	Barrett's esophagus	
	Head cancers	
	Brain cancer	
	Neck cancer	
	Breast cancer	
5-Aminolevulinic acid	Actinic keratosis	[54, 55]
	Basal cell carcinoma	
	Bladder tumor skin cancer	
	Lung cancer	
	Gastrointestinal tract	
Methyl aminolevulinate	Actinic keratosis	[56, 57]
Tetra(*m*-hydroxyphenyl)chlorin	Neck cancer	[58, 59]
	Scalp cancer	
	Breast cancer	
	Prostate cancer	
	Pancreatic cancers	
N-Aspartyl chlorin e6	Fibrosarcoma	[53]
	Liver cancer	
	Brain cancer	
	Oral cancer	
	Lung cancer	

[50]. Moreover, the application of hypericin has been studied in the clinical stage for PDT of basal and squamous cell carcinoma as well as mesothelioma. Further application of hypericin in a clinical-stage has been limited due to lack of solubility in aqueous media leading to immediate aggregations in water. Several studies have been improved the solubility of hypericin by using nanotechnology [51]. Lima et al. demonstrated that by entrapping hypericin in lipid-based nanoparticles, cellular uptake increased by 30% [51]. Table 6.2 summarized photosensitizers for clinical used.

Drug-delivery system represents an emerging approach to improve therapeutic outcomes of PDT and PTT. In addition, drug delivery systems provide the opportunity for dual cancer therapy by delivery both PTT or PDT agent and anticancer drugs in a single platform.

6.4   **NIR-triggered anticancer drug delivery**

Conventional cancer therapies have suffered from lack of specificity and toxicity. To solve the issue, nanotechnology-based drug delivery systems have been introduced to improve ERP effect, reduce toxicity and increase half-life during blood circulation. However, insufficient drug release in the tumorogenic side is the main drawback of using these systems. Stimuli-drug delivery systems have been provided the opportunity to tackle the limitation of typical drug delivery systems releasing the cargo under a specific stimulus which can be both internal or external. Although internal stimuli—such as pH, hypoxia, temperature, and so on—improved drug accumulation in tumor tissue, they have some limitations regarding poor control on drug release and difficulty in synthesis [60]. Compared to internal stimuli, external stimuli—such as light, magnetic field, ultrasound, and so on—release their cargo in a controllable manner during excitation [139]. Among these, NIR-light stimuli-responsive systems (NIRSRS) have absorbed great attention due to the ease of synthesis as compared to other stimuli-sensitive systems and deep penetration depth compared to other light-stimuli sensitive systems. There are three different mechanisms in NIRSRS for drug release included: (1) photothermal effect, (2) two-photon conversion, and (3) upconverting nanoparticles (UCNPs) [61]. These mechanisms are shown in Fig. 6.4.

6.4.1   **Photothermal-guided drug release (PT-NIRSRS)**

To prepare NIRSRS, thermo-sensitive materials are employed. These materials convert irradiated NIR light to heat which leads to an increase in the temperature and subsequently increases cargo release from the carrier either by phase transition mechanism or disrupting whole or part of the carrier. In addition, as already mentioned, heat has a cytotoxic effect on cancer cells, so these systems are considered for chemo-photothermal therapy.

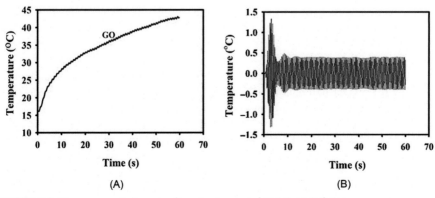

FIGURE 6.4 Three suggested mechanisms for laser-assisted drug delivery.

Being able to convert NIR light to heat, various materials have been reported include gold-based nanoparticles, carbon-based nanoparticles, metal-based nanoparticles, indocyanine green dye, polyaniline, melanin the main of which already described in Section 6.2.

In PT-NIRSRS, graphene oxide is more appropriate than graphene owing to the feasibility of drug-graphene oxide interaction and improve dispersibility. Also, surface modification of graphene oxide by polymers not only improves sustainability but also decrease cell toxicity as well. In 2017, multifunctional graphene oxide nanoparticles were synthesized to study the synergistic effect of chemo-gene and PTT by Zang et al. The PEGylated graphene oxides conjugated folate receptor was utilized to load both siRNA and doxorubicin. NIR irradiation not only exhibited cell toxicity but increased both gene and drug release as well [62]. In another study, graphene oxide was modified by polyglycerol to load curcumin [63]. Their nanoparticles exhibited excellent dispersibility, on-demand curcumin release during NIR irradiation and efficient chemo-photothermal therapy on MCF7. Owing to the presence of carboxylic and hydroxylic groups on its surface, graphene oxide has the potential to apply for multisensitive systems. In 2017, graphene oxide functionalized chitosan-PEG was synthesized to prepare both pH and NIR responsive system [64]. Due to its flexibility, graphene oxide is used as a coating. The nanohybrid composition of liposome and graphene oxide was prepared by layer by layer deposition of GO and graphene conjugated poly(L-lysine) (GO-PLL) on the surface of cationic liposome guided by electrostatic interaction. When the nanohybrid exposed to NIR laser irradiation, GO and GO-PLL functioned as a photothermal transducer and convert light to heat. Then, it activated a solid-to-gel phase transition of the liposomes leading to release the encapsulated toxic cargo. Their study introduced a novel pH-Thermo responsive system for chemo-photothermal therapy [65]. The chemo-photothermal potential effect of their nanoparticles was observed using live-dead assay under inverted fluorescent microscopy, as shown in Fig. 6.5. Live cells displayed in green fluorescent and dead cells displayed in red fluorescent. In the group treated by their nanoparticles (named LBL Lipo-graph), before NIR laser irradiation the dead cells had to do with chemotropic effect while after laser irradiation it was the correspondence of both chemo and photothermal effect. As shown in Fig. 6.5, in the cells treated by LBL Lipo-graph, dead cells increased after laser irradiation while in the control group (treated with no nanoparticles) the number of red spots had not changed before and after laser irradiation.

Another remarkable nanoparticle in PT-NIRSRS which already explained in Section 6.2.5 is carbon nanotubes. Although carbon nanotubes enjoy high cross-section absorption, high drug binding affinity and high ability to penetrate into cells, they suffer from low dispersibility and cell toxicity. To tackle the problem, Dong et al. conjugated TAT-chitosan to carbon nanotubes, and then doxorubicin was loaded on its surface. Results have shown that their nanoparticles exhibited NIR responsive properties during laser irradiation, pH-responsive behavior in the acidic environment, improving cell internalization due to the presence of chitosan [66].

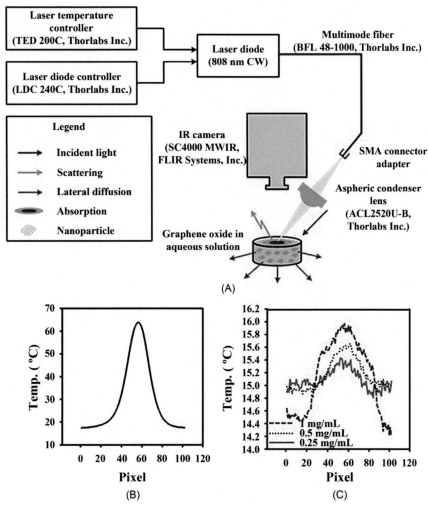

FIGURE 6.5 Live/death assay of LBL Lipo-graph fluorescent microscopy (the scale bar is 200 μm) [65].

Having high cross-section absorption and feasibility to drug binding, gold-based nanoparticles absorbed a great deal of attention in PT-NIRSRS. For example, gold nanoparticles and doxorubicin were embedded in pores of mesoporous silica. Then the chemo-photothermal therapy potentials of the carriers were examined on A549 cells. Their results demonstrated enhancing drug release in the presence of GSH and light irradiation [67]. In another study, the chemo-photothermal effect of GNR embedded in hydrogel was examined by Qu et al. Their thermoresponsive hydrogel was composed of N-isopropylacrylamide loaded with doxorubicin and GNRs. Their in vivo studies in 4T1 cells tumor-bearing mice showed a significant reduction in tumor size after NIR laser irradiation [68].

6.4.2 Two-photon conversion guided drug release (TP-NIRSRS)

Designing NIRSRS should have some characteristic features including strong absorption in NIR region, efficient photothermal conversion, efficient drug accumulation, and low toxicity. One of the main drawbacks of light-responsive drug release systems is that many of these systems require high energy wavelength of UV or visible light which have sufficient energy for activation.

As already discussed, UV light suffered from poor penetrating ability and harmful cell toxicity. To overcome this issue, two-photon absorption nanoparticles have been introduced characterized by absorption of two low-energy of NIR light and converting to high-energy visible light. This high visible light energy can sensitize oxygen to produce singlet oxygen or reactive oxygen species to kill abnormal cells. NIR light used to excite the nanoparticles lead to penetrating light to deep-seated tumors. Furthermore, two-photon absorption nanoparticles normally possess high cross-section absorption which has to do with efficient PDT therapy. Croissant et al. found that hydrophobic poly(2-nitrobenzyl methacrylate) and hydrophilic polyethylene oxide had two-photon conversion potential [69] which can be used for TP-NIRSRS. In another study, novel paracyclophane-based fluorophore nanoparticles were embedded on the pores of mesoporous silica. The fluorophore could absorb two-photon in NIR region and convert energy via FRET to azobenzene of valve leading to open valves through *cis-trans* isomerization and releasing encapsulated drug [70]. The results demonstrated an efficient cell death after NIR laser irradiation, while no significant cell death was observed without irradiation. In 2017, novel nitrogen-doped carbon dots were synthesized for TP-NIRSRS. Ardekaniet et al. found that their nanoparticles exhibited both NIR responsive drug release and photothermal cell toxicity in in vitro model [71].

6.4.3 Upconverting nanoparticles guided drug release (UP-NIRSRS)

UCNPs can convert NIR light to high-energy UV and visible light mainly made of lanthanides. Then, this high energy can activate photochemical reactions leading to an increase in cargo release or activate photosensitizers for PDT application. UCNPs nanoparticles have improved the main drawback of UV sensitive nanoparticle, that is, toxicity and poor penetration dept. Three ways have been studied to fabricate UP- NIRSRS: (a) UCNPs and drug have been encapsulated in hydrophobic carriers via hydrophobic-hydrophobic interaction. (b) Mesoporous silica-coated UCNPs are providing large surface area for drug deposition. (c) Encapsulating of drug and UCNPs nanoparticles in spherical carriers [71]. In 2013, Liu et al. introduced novel UP- NIRSRS composed of NaYF4:TmYb core-shell with a coating layer of mesoporous silica containing azobenzene groups. The cellular uptake was further improved using TAT protein. When the nanoparticles exposed to NIR laser irradiation, UCNPs convert NIR light to UV. UV emission led to transfer azobenzene to *trans* configuration had to do with releasing of encapsulated cargo [72]. In another study, Hu et al. developed a novel UP-NIRSRS composed of NaYF4:Yb:Er coated with octadecyl-quaternized polyglutamic acid and polystyrene-block-poly(acrylic acid) which further modified by zinc(II)phthalocyanine as a photosensitizer [73]. Similarly, in another study, nanoimpellers composed of NaYF4:Yb and Er covered with mesoporous silica were developed for chemo-PDT therapy. The cytotoxicity results displayed a synergistic effect of chemotherapy and PDT [74].

6.5   Cross-section absorption of graphene oxide

Besides the mathematical models described in Section 6.1, there are several experiments to study optical properties of the materials. In 2018, Hashemi et al. obtained the cross-section absorption of graphene oxide by calculating absorption coefficient of the sample volume and normalized it to the graphene oxide concentration. When laser emitted to the materials, the energy can be either reflect from the surface, diffused via lateral parts of the sample holder, scatter or absorbed by the nanoparticles. To calculate cross-section absorption, at first step heat loss trough lateral part of the sample holder and scattering need to be minimized [75].

6.5.1   Minimization of lateral thermal diffusion and scattering

Lateral thermal diffusion was minimized by calculation of lateral thermal relaxation:

$$\tau = \frac{d^2}{16x} \tag{6.19}$$

where d is the laser beam diameter (2 mm) and x is the thermal diffusivity of water (0.14 mm^2 s^{-1}). Based on the equation above, the relaxation time was 1.78 s. So, by choosing laser pulse with reputation rate of 2 Hz (50% duty cycle), the pulse duration would be well below than lateral thermal relaxation time.

Scattering was minimized by comparing the temperature profile of various concentrations of GO (0.25, 0.125, and 0.0625 mg mL^{-1}) with the temperature profile of standard black absorber (Fig. 6.6). The concentration of 0.25 mg/mL which has a similar temperature profile spread to the black absorber was chosen as a concentration with minimum scattering effect.

6.5.2   Obtaining cross-section absorption of GO

By reducing the effect of scattering and heat loss through the lateral part of sample holder, the heat generated during pulsed laser irradiation is almost absorbed by the nanoparticles:

$$\frac{dQ}{dt} = mC\frac{dT}{dt} \tag{6.20}$$

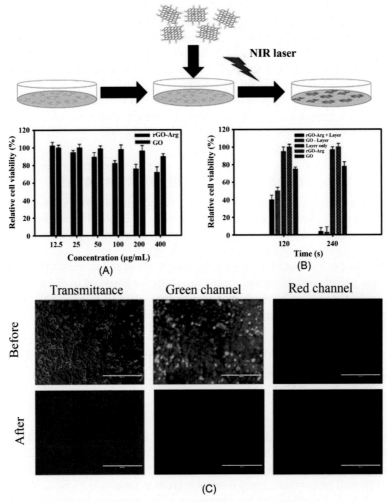

FIGURE 6.6 Schematic of set up to measure absorption cross-section.

Scattering profile of (GO at different concentration, (C) black absorber [75].

where Q is the heat generated during laser emission (J), m is the mass of the nanoparticle (kg)], C is the specific heat capacity (J/kg K). By substituting $\frac{dQ/dt}{V} = \mu_a \times \varphi$ and $m/V = \rho$ in the equation above:

$$\frac{dT(t)}{dt} = \frac{\mu_a \times \varphi(t)}{\rho C} \qquad (6.21)$$

where μ_a is the absorption coefficient (L mol^{-1} cm^{-1}), φ is laser beam fluency (W/m^2), ρ is material density (kg/m^3). By substituting analytic signals for laser fluency and

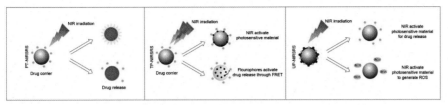

FIGURE 6.7

(A) Temporal profile of temperature versus time of GO and (B) filtering of Plat a.

temperature $\varphi = \varphi_0 e^{i2\pi ft}$ and $T = T_0 e^{i2\pi ft}$ in the equation above, the magnitude of absorption coefficient is:

$$\mu_a = \frac{2\pi f \rho C T_0}{\varphi_o} \tag{6.22}$$

So, the absorption cross-section of the graphene oxide is calculated using the following equation:

$$\mu_a = \sigma * C \tag{6.23}$$

where σ is absorption cross-section (mL/mg m), and C is the corresponding concentration (mg/mL).

Temperature per pulse, T_0, is calculated by placing 400 μL of the graphene oxide in 96 well-plate, irradiating with a 808 nm NIR laser pulse (2 Hz at 50% duty cycle) and monitoring temperature rise for a minute. Then the measured temporal profile is filtered with a bandpass filter centered at 2 Hz, to make sure that the slope is not affected by intensity noise (Fig. 6.7). The steady-state temperature per pulse for GO was 0.16 ± 0.01 °C. By substituting $\varphi_0 = 1.77 \,(\text{W/cm}^2)$, $\rho = 1001.77 \,(\text{kg/m}^3)$, $C = 4178.62$ (J/kg K), and $f = 2\,\text{Hz}$, the absorption cross-section of GO was 19 ± 1.2 (mL(mg cm)^{-1}).

Laser-assisted cancer therapy is a highly efficient and minimally invasive method with the ability to absorb incident light and convert it to heat or singlet oxygen to destroy cancer cells. Several organic and inorganic materials have been investigated for laser-assisted cancer therapy. PDT and PTT have been successfully applied for cancer therapy due to several advantages including inducing cancer cells to apoptosis as well as sensitize cancer cells to chemotherapy, immunotherapy and gene therapy. An ideal photothermal or photodynamic transducers must meet several requirements including having a tunable light-matter response, having high cross-section absorption, having deep penetration depth, needing a simple modification process and having an efficient tumor accumulation. Furthermore, to improve therapeutic outcomes, a combination of other imaging and treatment approaches have been investigated. Also, administrating photosensitive agent in drug delivery vehicles provide the opportunity for fabricating stimuli-sensitive drug delivery systems which have the ability to increase localize delivery. A multifunctional system integrating into a single platform has a potential application in a large verity of diseases. Several mathematical and experimental studies focused on investigating the optical properties of nanoparticles.

Reference

[1] G. Mie, Beiträge zur Optik trüber Medien, speziell kolloidaler Metallösungen, Ann. Phys. 330 (3) (1908) 377–445.

[2] C. Mätzler, MATLAB functions for Mie scattering and absorption, version 2, IAP Res. Rep. 8 (1) (2002) 9.

[3] C.D. Mobley, Radiative Transfer Modeling for CoBOP, Sequoia Scientific Inc., Redmond WA WestPark Technical Center, 2001.

[4] B.T. Draine, P.J. Flatau, Discrete-dipole approximation for scattering calculations, JOSA A 11 (4) (1994) 1491–1499.

[5] J.G. Morelli, O.T. Tan, R. Margolis, Y. Seki, J. Boll, J.M. Carney, et al. Tunable dye laser (577 nm) treatment of port wine stains, Lasers Surg. Med. 6 (1) (1986) 94–99.

[6] W.R. Chen, R.L. Adams, S. Heaton, D.T. Dickey, K.E. Bartels, R.E. Nordquist, Chromophore-enhanced laser-tumor tissue photothermal interaction using an 808-nm diode laser, Cancer Lett. 88 (1) (1995) 15–19.

[7] H. Takahashi, T. Niidome, A. Nariai, Y. Niidome, S. Yamada, Gold nanorod-sensitized cell death: microscopic observation of single living cells irradiated by pulsed near-infrared laser light in the presence of gold nanorods, Chem. Lett. 35 (5) (2006) 500–501.

[8] H. Golzar, F. Yazdian, M. Hashemi, M. Omidi, D. Mohammadrezaei, H. Rashedi, et al. Optimizing the hybrid nanostructure of functionalized reduced graphene oxide/silver for highly efficient cancer nanotherapy, New J. Chem. 42 (15) (2018) 13157–13168.

[9] S.J. Oldenburg, R.D. Averitt, S.L. Westcott, N.J. Halas, Nanoengineering of optical resonances, Chem. Phys. Lett. 288 (2–4) (1998) 243–247.

[10] N. Biosciences, Inc. Pilot Study of Aurolase Therapy in Refractory and/or Recurrent Tumors of the Head and Neck, ClinicalTrials. gov [Internet], National Library of Medicine, Bethesda, MD, USA, 2000.

[11] L.R. Hirsch, R.J. Stafford, J.A. Bankson, S.R. Sershen, B. Rivera, R.E. Price, et al. Nanoshell-mediated near-infrared thermal therapy of tumors under magnetic resonance guidance, Proc. Natl. Acad. Sci. 100 (23) (2003) 13549–13554.

[12] C. Loo, A. Lowery, N. Halas, J. West, R. Drezek, Immunotargeted nanoshells for integrated cancer imaging and therapy, Nano Lett. 5 (4) (2005) 709–711.

[13] R. Fekrazad, N. Hakimiha, E. Farokhi, M.J. Rasaee, M.S. Ardestani, K.A. Kalhori, F. Sheikholeslami, Treatment of oral squamous cell carcinoma using anti-HER2 immunonanoshells, Int. J. Nanomed. 6 (2011) 2749.

[14] S.Y. Liu, Z.S. Liang, F. Gao, S.F. Luo, G.Q. Lu, In vitro photothermal study of gold nanoshells functionalized with small targeting peptides to liver cancer cells, J. Mater. Sci.: Mater. Med. 21 (2) (2010) 665–674.

[15] M.P. Melancon, W. Lu, M. Zhong, M. Zhou, G. Liang, A.M. Elliott, et al. Targeted multifunctional gold-based nanoshells for magnetic resonance-guided laser ablation of head and neck cancer, Biomaterials 32 (30) (2011) 7600–7608.

[16] X. Huang, I.H. El-Sayed, W. Qian, M.A. El-Sayed, Cancer cell imaging and photothermal therapy in the near-infrared region by using gold nanorods, J. Am. Chem. Soc. 128 (6) (2006) 2115–2120.

[17] E.B. Dickerson, E.C. Dreaden, X. Huang, I.H. El-Sayed, H. Chu, S. Pushpanketh, et al. Gold nanorod assisted near-infrared plasmonic photothermal therapy (PPTT) of squamous cell carcinoma in mice, Cancer Lett. 269 (1) (2008) 57–66.

[18] D. Pissuwan, S.M. Valenzuela, M.C. Killingsworth, X. Xu, M.B. Cortie, Targeted destruction of murine macrophage cells with bioconjugated gold nanorods, J. Nanopart. Res. 9 (6) (2007) 1109–1124.

[19] D. Pissuwan, S.M. Valenzuela, C.M. Miller, M.B. Cortie, A golden bullet? Selective targeting of Toxoplasma gondii tachyzoites using antibody-functionalized gold nanorods, Nano Lett. 7 (12) (2007) 3808–3812.

[20] G. Von Maltzahn, J.H. Park, A. Agrawal, N.K. Bandaru, S.K. Das, M.J. Sailor, S.N. Bhatia, Computationally guided photothermal tumor therapy using long-circulating gold nanorod antennas, Cancer Res. 69 (9) (2009) 3892–3900.

[21] K.C. Black, J. Yi, J.G. Rivera, D.C. Zelasko-Leon, P.B. Messersmith, Polydopamine-enabled surface functionalization of gold nanorods for cancer cell-targeted imaging and photothermal therapy, Nanomedicine 8 (1) (2013) 17–28.

[22] Y. Akiyama, T. Mori, Y. Katayama, T. Niidome, The effects of PEG grafting level and injection dose on gold nanorod biodistribution in the tumor-bearing mice, J. Control. Release 139 (1) (2009) 81–84.

[23] Y.S. Chen, W. Frey, S. Kim, K. Homan, P. Kruizinga, K. Sokolov, S. Emelianov, Enhanced thermal stability of silica-coated gold nanorods for photoacoustic imaging and image-guided therapy, Optics express 18 (9) (2010) 8867–8878.

[24] Y. Qiu, Y. Liu, L. Wang, L. Xu, R. Bai, Y. Ji, et al. Surface chemistry and aspect ratio mediated cellular uptake of Au nanorods, Biomaterials 31 (30) (2010) 7606–7619.

[25] J. Chen, D. Wang, J. Xi, L. Au, A. Siekkinen, A. Warsen, et al. Immuno gold nanocages with tailored optical properties for targeted photothermal destruction of cancer cells, Nano Lett. 7 (5) (2007) 1318–1322.

[26] L. Au, D. Zheng, F. Zhou, Z.Y. Li, X. Li, Y. Xia, A quantitative study on the photothermal effect of immuno gold nanocages targeted to breast cancer cells, ACS Nano 2 (8) (2008) 1645–1652.

[27] J. Chen, C. Glaus, R. Laforest, Q. Zhang, M. Yang, M. Gidding, et al. Gold nanocages as photothermal transducers for cancer treatment, Small 6 (7) (2010) 811–817.

[28] S.A. Khan, R. Kanchanapally, Z. Fan, L. Beqa, A.K. Singh, D. Senapati, P.C. Ray, A gold nanocage–CNT hybrid for targeted imaging and photothermal destruction of cancer cells, Chem. Commun. 48 (53) (2012) 6711–6713.

[29] B. Khlebtsov, E. Panfilova, V. Khanadeev, O. Bibikova, G. Terentyuk, A. Ivanov, et al. Nanocomposites containing silica-coated gold–silver nanocages and Yb–2, 4-dimethoxy-hematoporphyrin: multifunctional capability of IR-luminescence detection, photosensitization, and photothermolysis, ACS Nano 5 (9) (2011) 7077–7089.

[30] L. Gao, J. Fei, J. Zhao, H. Li, Y. Cui, J. Li, Hypocrellin-loaded gold nanocages with high two-photon efficiency for photothermal/photodynamic cancer therapy in vitro, ACS Nano 6 (9) (2012) 8030–8040.

[31] S.K. Dondapati, T.K. Sau, C. Hrelescu, T.A. Klar, F.D. Stefani, J. Feldmann, Label-free biosensing based on single gold nanostars as plasmonic transducers, ACS Nano 4 (11) (2010) 6318–6322.

[32] H. Yuan, C.G. Khoury, C.M. Wilson, G.A. Grant, A.J. Bennett, T. Vo-Dinh, In vivo particle tracking and photothermal ablation using plasmon-resonant gold nanostars, Nanomed.: Nanotechno. Biol. Med. 8 (8) (2012) 1355–1363.

[33] H. Yuan, A.M. Fales, T. Vo-Dinh, TAT peptide-functionalized gold nanostars: enhanced intracellular delivery and efficient NIR photothermal therapy using ultralow irradiance, J. Am. Chem. Soc. 134 (28) (2012) 11358–11361.

[34] H.K. Moon, S.H. Lee, H.C. Choi, In vivo near-infrared mediated tumor destruction by photothermal effect of carbon nanotubes, ACS Nano 3 (11) (2009) 3707–3713.

[35] J.T. Robinson, G. Hong, Y. Liang, B. Zhang, O.K. Yaghi, H. Dai, In vivo fluorescence imaging in the second near-infrared window with long circulating carbon nanotubes capable of ultrahigh tumor uptake, J. Am. Chem. Soc. 134 (25) (2012) 10664–10669.

[36] A.L. Antaris, J.T. Robinson, O.K. Yaghi, G. Hong, S. Diao, R. Luong, H. Dai, Ultra-low doses of chirality sorted (6.5) carbon nanotubes for simultaneous tumor imaging and photothermal therapy, ACS Nano 7 (4) (2013) 3644–3652.

[37] K. Yang, S. Zhang, G. Zhang, X. Sun, S.T. Lee, Z. Liu, Graphene in mice: ultrahigh in vivo tumor uptake and efficient photothermal therapy, Nano Lett. 10 (9) (2010) 3318–3323.

[38] J.T. Robinson, S.M. Tabakman, Y. Liang, H. Wang, H. Sanchez Casalongue, D. Vinh, H. Dai, Ultrasmall reduced graphene oxide with high near-infrared absorbance for photothermal therapy, J. Am. Chem. Soc. 133 (17) (2011) 6825–6831.

[39] M. Hashemi, M. Omidi, B. Muralidharan, H. Smyth, M.A. Mohagheghi, J. Mohammadi, T.E. Milner, Evaluation of the photothermal properties of a reduced graphene oxide/arginine nanostructure for near-infrared absorption, ACS Appl. Mater Interfaces 9 (38) (2017) 32607–32620.

[40] O. Raab, On the effect of fluorescent substances on infusoria, Z. Biol. 39 (1900) 524–526.

[41] N.A. Samy, M.M. Salah, M.F. Ali, A.M. Sadek, Effect of methylene blue-mediated photodynamic therapy for treatment of basal cell carcinoma, Lasers Med. Sci. 30 (1) (2015) 109–115.

[42] A.R. Disanto, J.G. Wagner, Pharmacokinetics of highly ionized drugs II: methylene blue—absorption, metabolism, and excretion in man and dog after oral administration, J. Pharmaceut. Sci. 61 (7) (1972) 1086–1090.

[43] A.E. da Hora Machado, Terapia fotodinâmica: princípios, potencial de aplicação e perspectivas, Química Nova 23 (2) (2000).

[44] J. Ferreira, P.F.C. Menezes, C.H. Sibata, R.R. Allison, S. Zucoloto, O.C. e Silva, V.S. Bagnato, Can efficiency of the photosensitizer be predicted by its photostability in solution?, Laser Phys. 19 (9) (2009) 1932–1938.

[45] A. Parihar, A. Dube, P.K. Gupta, Conjugation of chlorin 6 to histamine enhances its cellular uptake and phototoxicity in oral cancer cells, Cancer Chemother. Pharmacol. 68 (2) (2011) 359–369.

[46] M.M. LoTempio, M.S. Veena, H.L. Steele, B. Ramamurthy, T.S. Ramalingam, A.N. Cohen, Curcumin suppresses growth of head and neck squamous cell carcinoma, Clin. Cancer Res. 11 (19) (2005) 6994–7002.

[47] M. Mohan Yallapu, M. Ray Dobberpuhl, D. Michele Maher, M. Jaggi, S. Chand Chauhan, Design of curcumin loaded cellulose nanoparticles for prostate cancer, Curr. Drug Metab. 13 (1) (2012) 120–128.

[48] L.A. Muehlmann, B.C. Ma, J.P.F. Longo, M.D.F.M.A. Santos, R.B. Azevedo, Aluminum–phthalocyanine chloride associated to poly(methyl vinyl ether-co-maleic anhydride) nanoparticles as a new third-generation photosensitizer for anticancer photodynamic therapy, Int. J. Nanomed. 9 (2014) 1199.

[49] B. Kleemann, B. Loos, T.J. Scriba, D. Lang, L.M. Davids, St John's Wort (Hypericum perforatum L.) photomedicine: hypericin-photodynamic therapy induces metastatic melanoma cell death, PLoS ONE 9 (7) (2014) e103762.

[50] M. Barathan, V. Mariappan, E.M. Shankar, B.J. Abdullah, K.L. Goh, J. Vadivelu, Hypericin-photodynamic therapy leads to interleukin-6 secretion by HepG2 cells and

their apoptosis via recruitment of BH3 interacting-domain death agonist and caspases, Cell Death Dis. 4 (6) (2013) e697.

[51] A.M. Lima, C. Dal Pizzol, F.B. Monteiro, T.B. Creczynski-Pasa, G.P. Andrade, A.O. Ribeiro, J.R. Perussi, Hypericin encapsulated in solid lipid nanoparticles: phototoxicity and photodynamic efficiency, J. Photochem. Photobiol. B: Biol. 125 (2013) 146–154.

[52] S.K. Pushpan, S. Venkatraman, V.G. Anand, J. Sankar, D. Parmeswaran, S. Ganesan, T.K. Chandrashekar, Porphyrins in photodynamic therapy-a search for ideal photosensitizers, Curr. Med. Chem.-Anti-Cancer Agents 2 (2) (2002) 187–207.

[53] J. Usuda, H. Kato, T. Okunaka, K. Furukawa, H. Tsutsui, K. Yamada, et al. Photodynamic therapy (PDT) for lung cancers, J. Thoracic Oncol. 1 (5) (2006) 489–493.

[54] T.J. Dougherty, An update on photodynamic therapy applications, J. Clin. Laser Med. Surg. 20 (1) (2002) 3–7.

[55] C. Morton, S.B. Brown, S. Collins, S. Ibbotson, H. Jenkinson, H. Kurwa, et al. Guidelines for topical photodynamic therapy: report of a workshop of the British Photodermatology Group, Br. J. Dermatol. 146 (4) (2002) 552–567.

[56] J.W. Lee, H.I. Lee, M.N. Kim, B.J. Kim, Y.J. Chun, D. Kim, Topical photodynamic therapy with methyl aminolevulinate may be an alternative therapeutic option for the recalcitrant Malassezia folliculitis, Int. J. Dermatol. 50 (4) (2011) 488–490.

[57] C.A. Morton, Methyl aminolevulinate: actinic keratoses and Bowen's disease, Dermatol. Clinics 25 (1) (2007) 81–87.

[58] M.O. Senge, J.C. Brandt, Temoporfin (Foscan®, 5, 10, 15, 20-tetra (m-hydroxyphenyl) chlorin)—a second-generation photosensitizer, Photochemistry and photobiology 87 (6) (2011) 1240–1296.

[59] M. Triesscheijn, M. Ruevekamp, M. Aalders, P. Baas, F.A. Stewart, Outcome of mTHPC mediated photodynamic therapy is primarily determined by the vascular response, Photochem. Photobiol. 81 (5) (2005) 1161–1167.

[60] C. Hang, Y. Zou, Y. Zhong, Z. Zhong, F. Meng, NIR and UV-responsive degradable hyaluronic acid nanogels for CD44-targeted and remotely triggered intracellular doxorubicin delivery, Coll. Surf. B: Biointerfaces 158 (2017) 547–555.

[61] A. Raza, U. Hayat, T. Rasheed, M. Bilal, H.M. Iqbal, Smart materials-based near-infrared light-responsive drug delivery systems for cancer treatment: a review, J. Mater. Res. Technol. 8 (1) (2019) 1497–1509.

[62] Y. Zeng, Z. Yang, H. Li, Y. Hao, C. Liu, L. Zhu, et al. Multifunctional nanographene oxide for targeted gene-mediated thermochemotherapy of drug-resistant tumour, Sci. Rep. 7 (2017) 43506.

[63] F. Bani, M. Adeli, S. Movahedi, M. Sadeghizadeh, Graphene–polyglycerol–curcumin hybrid as a near-infrared (NIR) laser stimuli-responsive system for chemo-photothermal cancer therapy, RSC Adv. 6 (66) (2016) 61141–61149.

[64] R.K. Thapa, J.H. Byeon, S.K. Ku, C.S. Yong, J.O. Kim, Easy on-demand self-assembly of lateral nanodimensional hybrid graphene oxide flakes for near-infrared-induced chemothermal therapy, NPG Asia Mater. 9 (8) (2017) e416.

[65] M. Hashemi, M. Omidi, B. Muralidharan, L. Tayebi, M.J. Herpin, M.A. Mohagheghi, et al. Layer-by-layer assembly of graphene oxide on thermosensitive liposomes for photo-chemotherapy, Acta Biomater. 65 (2018) 376–392.

[66] X. Dong, Z. Sun, X. Wang, X. Leng, An innovative MWCNTs/DOX/TC nanosystem for chemo-photothermal combination therapy of cancer, Nanomed.: Nanotechnol. Biol. Med. 13 (7) (2017) 2271–2280.

[67] Y. Yang, Y. Lin, D. Di, X. Zhang, D. Wang, Q. Zhao, S. Wang, Gold nanoparticle-gated mesoporous silica as redox-triggered drug delivery for chemo-photothermal synergistic therapy, J. Coll. Interface Sci. 508 (2017) 323–331.

[68] Y. Qu, B.Y. Chu, J.R. Peng, J.F. Liao, T.T. Qi, K. Shi, et al. A biodegradable thermo-responsive hybrid hydrogel: therapeutic applications in preventing the post-operative recurrence of breast cancer, NPG Asia Mater. 7 (8) (2015) e207.

[69] J. Croissant, A. Chaix, O. Mongin, M. Wang, S. Clément, L. Raehm, et al. Two-photon-triggered drug delivery via fluorescent nanovalves, Small 10 (9) (2014) 1752–1755.

[70] S.M. Ardekani, A. Dehghani, M. Hassan, M. Kianinia, I. Aharonovich, V.G. Gomes, Two-photon excitation triggers combined chemo-photothermal therapy via doped carbon nanohybrid dots for effective breast cancer treatment, Chem. Eng. J. 330 (2017) 651–662.

[71] G. Chen, H. Qiu, P.N. Prasad, X. Chen, Upconversion nanoparticles: design, nanochemistry, and applications in theranostics, Chem. Rev. 114 (10) (2014) 5161–5214.

[72] J. Liu, W. Bu, L. Pan, J. Shi, NIR-triggered anticancer drug delivery by upconverting nanoparticles with integrated azobenzene-modified mesoporous silica, Angew. Chem. Int. Ed. 52 (16) (2013) 4375–4379.

[73] B. Hou, B. Zheng, W. Yang, C. Dong, H. Wang, J. Chang, Construction of near infrared light triggered nanodumbbell for cancer photodynamic therapy, J. Coll. Interface Sci. 494 (2017) 363–372.

[74] B. Hou, W. Yang, C. Dong, B. Zheng, Y. Zhang, J. Wu, et al. Controlled co-release of doxorubicin and reactive oxygen species for synergistic therapy by NIR remote-triggered nanoimpellers, Mater. Sci. Eng. C 74 (2017) 94–102.

[75] M. Hashemi, B. Muralidharan, M. Omidi, J. Mohammadi, Y. Sefidbakht, E.S. Kim, et al. Effect of size and chemical composition of graphene oxide nanoparticles on optical absorption cross-section, J. Biomed. Opt. 23 (8) (2018) 085007.

Application of magnetic and electric fields for cancer therapy

Chapter outline

7.1 Magnetic nanoparticles properties

Nano drug carriers are the newest tools in drug delivery. Nanocarriers size is 1–100 nm and can deliver drugs to sites that are otherwise inaccessible in the human body. They are classified based on material, shape, and different medical applications [1]. In recent years the magnetic nanoparticles (MNPs) is used as drug carrier for human therapy. MNPs, such as iron and oxides nanoparticles, are nano-sized particles made of materials with magnetic properties. The very small diameter and magnetic properties make the MNPs suitable for medical applications such as targeted delivery. The ability of MNPs to accept different surface coatings is among the other distinct feature of MNPs biomedical properties. An example of this feature is the stability of the therapeutic carriers in human body. Another feature of MNPs is the ability to interact with cells or biological proteins to avoid toxicity and to increase the biocompatibility. MNPs appreciable properties with the magnetic field make them intelligent particles that can be guided remotely by changing magnet gradient and intensity [1]. The composition, size, and pathway of MNP synthesis vary according to their application. In addition, different types of MNPs are designed for use in specific applications.

Magnetization is a base of classification for magnetic materials. Accordingly, the materials classify into three groups: ferromagnetic, paramagnetic, and diamagnetic.

Magnetization of diamagnetic materials is negative and very low. The resultant magnetic dipole moment in these materials is zero [1]. Pankhurst et al. discussed the laws of the magnetic field, the various uses of MNPs in medicine and identified the permitted magnetic fields for different applications [2]. Single domain ferromagnetic nanoparticles and superparamagnetic nanoparticles are the two important MNPs features which are elaborated future.

Single domain ferromagnetic nanoparticles: A magnetic particle that stays in a single domain state for all magnetic fields is called a single-domain ferromagnetic particle. Usually nanoparticles with a diameter of less than 100 nm are single-domain nanoparticles [3,4]. The internal magnetic moment of the single-domain ferromagnetic nanoparticles is not zero.

Superparamagnetic nanoparticles: The effective time constant for nanoparticles smaller than the 10–15 nm is very small. This causes rapid fluctuations, the microsecond period, in the magnetic field. Therefore, in the absence of external magnetic field, the magnetic moment of the nanoparticle and the internal magnetic moment is zero during this very short period. This property is called superparamagnetism [5]. The superparamagnetism expression arises from the coupling of many atomic spins and is used to indicate an analogy between the behavior of the small magnetic moment of a single paramagnetic atom and that of the much larger magnetic moment of a MNPs [5]. In the presence of the external magnetic field, superparamagnetism nanoparticles are rapidly aligned with the field. The maximum diameter of the nanoparticles with superparamagnetic properties (D_{SPM}) is obtained from the following equation [4,6]:

$$D_{SPM} = 2\sqrt[3]{\frac{6k_B T}{K}} \tag{7.1}$$

MNPs that used in medical applications belong to the paramagnetic materials [1]. Fig. 7.1 depicts several types of drug carriers with MNPs.

The coating of magnetic nanocarriers is designed in a manner that it can accomplish different mechanisms. Fig. 7.2 schematically shows a core-shell nanocarrier with several factors on its surface.

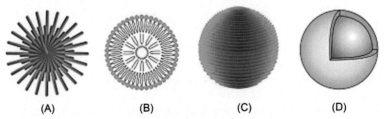

(A)	(B)	(C)	(D)

FIGURE 7.1 Schematic structure of magnetic nanoparticles and their coatings.

(A) End-grafted polymer-coated MNP, (B) liposome-encapsulated MNP, (C) fully encapsulated in polymer coating MNP, and (D) core-shell MN.

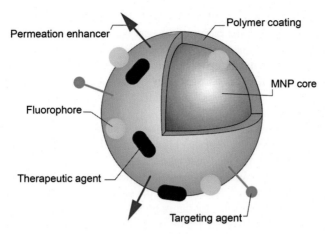

FIGURE 7.2 Schematic of the agents designed on the surface of a core-shell MNP.

The simplified model of a MNP with a polymer core for drug delivery is depicted in Fig. 7.3. As shown, the core of the carrier is made up of polymer materials that are placed inside the magnetic nanocarrier.

Sun et al. [7] and Chomoucka et al. [8] explained in detail the data related to nanocarriers type and their coatings.

Nanoparticles carriers for drug delivery and hyperthermia application must obtain the following characteristics [9]:

Nontoxic characteristic: The nanocarriers based on iron oxide are less toxic than nickel and cobalt nanocarriers. The toxicity may occur after injection of more than 250 mg of iron per kilogram of the mouse body [5].

Proper dimension characteristic: The nanocarriers size, as their name implies, must be small enough for easy penetration into the cancerous tumors. They should also be large enough to overcome the surface forces that are caused by the blood flow. Otherwise these forces dominate the flow and prevent the focused delivery of the

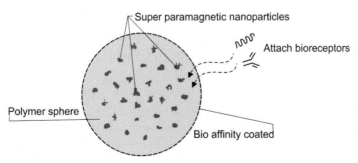

FIGURE 7.3 Schematic of a polymer core drug carrier with magnetic nanoparticles.

nanocarrier to the desired part of the tissue [10]. This is because the surface force has a linear relationship with the radius of the nanocarriers and the magnetic force varies with the third power of the nanocarriers radius.

High magnetism characteristic: Nanocarriers with high magnetism are suitable for precise delivery of the nanocarriers to the targeted tissue [9]. Therefore, higher magnetism causes the hysteresis and residual losses of the nanocarriers, the main sources of heat production, to be high.

Long lifespan characteristic: When foreign substances enter the blood circulation system, the immune system quickly detects and eliminates them. This immune system also applies to injected nanoparticles and shortly detects and repels them. The duration of identification and repulsion depends on the size of the nanoparticles and their surface properties. By choosing the appropriate surface for nanoparticles, the absorption of nanoparticles by the immune system is delayed. In general, larger nanoparticles are eliminated earlier by the immune system [10]. Based on in vitro results, the resistance time in the blood plasma for a type of iron oxide nanoparticles (diameter of 100 nm) was reported to be about 30 min [10]. Gamarra et al. [11] determined experimentally the detection and elimination of iron oxide nanoparticles with dextran coatings in mice. Also, they measured the concentration of nanoparticles in the rat.

Hydrophobic characteristic: Hydrophobic nanocarriers are carriers that do not mix with water. Generally, the release of hydrophobic nanocarriers is more homogeneous in the blood plasma than other types of carriers. Suitable coatings of nanocarriers prevent them from oxidation and bonding. A great deal of research is focused on obtaining suitable nanoparticles for medical applications [9]. Dobrovolskaia et al. studied the effect of 12 types of PAMAM dendrimers coatings on the human platelets in vitro [12]. Regarding the above properties, polyethylene glycol molecule is most used to prevent the operation of the immune system. Of course, carbon and silica coatings are also other options for nanoparticle coatings that have their own properties [9].

Considering the required characteristics and the production cost, nanocarriers based on iron oxide is most suitable for medical applications [5,9]. Chertok et al. showed that iron oxide nanocarriers are also good for delivering the drug to the brain tumors [13].

7.2 Bioengineering application of electromagnetic fields

It is well known that electromagnetic fields along with electrical conductor materials lead to an increase in temperature by means of Joule effect or energy deposition. Also, they might induce the same effects on human body structures. Actually, the electromagnetic fields, with the same mechanisms, induce current density or heat in human body [14]. These interactions are used in medical treatments. To reach better therapeutic effect, human body tissue heating is used for hyperthermia, drug delivery as well as muscle contractions or nerve stimulation. Advanced research in electromagnetics theory has played a fundamental role in the development of bio-medical technology. These studies include evaluation of health hazards of microwave field emission with biological tissues and living systems, and also the therapeutic, diagnostic, and imaging applications of electromagnetics.

Some examples of electromagnetic nanoparticles that are widely used in medicine are:

- Promote hyperthermia therapy in cancer treatment
- Improved imaging quality based on magnetic resonance
- Separate cells and macromolecules and purifies cells
- Application in biomass sensors
- Help with the transfer of the desired composition including gene, drug, stem cell, protein, and antibody to the targeted tissue
- Help with detecting particles in the in vivo and in vitro conditions using magnetic resonance imaging (MRI)

In the following section the application of electromagnetic nanoparticles is presented.

Magnetic separation: One of the MNPs application is to separate the desired material/nanoparticles cells in the bio solution by utilizing a magnetic field and then to count their numbers. Initially the magnetic field identifies and removes the labeled MNPs from the main solution. Then the number of cells is calculated by measuring the magnetic properties of the separated material [5].

Improved imaging quality based on magnetic resonance: MRI is one of the relatively new methods for medical imaging. Some of the MRI advantages are its high contrast in soft tissue, proper resolution and high penetration depth for noninvasive clinical diagnosis. The main goal of the MRI is to increase the efficiency of the imaging technique, to enhance the contrast and thus to distinguish healthy tissues from the rest. For example, in order to enhance the contrast of the tissue images, a MNPs based on iron oxide that is nontoxin is injected into the tissue before using the MRI [5].

7.3 Governing equations

The governing equations related to electromagnetism are: The Gaussian equations for electric and magnetic fields, Faraday's law, Maxwell-Ampere equation, and the electric charge conservation and are as follows, respectively [15].

Gauss equation for electric field:

$$\nabla \cdot E = \frac{\rho}{\varepsilon_0} \tag{7.2}$$

Gauss equation for magnetic field:

$$\nabla \cdot B = 0 \tag{7.3}$$

Faraday's law:

$$\nabla \times E = -\frac{\partial B}{\partial t} \tag{7.4}$$

Maxwell-Ampere equation:

$$\nabla \times B = \mu_0 j + \frac{1}{C^2}\frac{\partial E}{\partial t} \tag{7.5}$$

Electric charge conservation:

$$\nabla \cdot j = -\frac{\partial \rho}{\partial t} \tag{7.6}$$

where ρ, E, B, j, ε_0, μ_0 and C are electric charge density, electrical field, magnetic field, electric current density, free space permittivity, vacuum magnetic permeability, and speed of light, respectively.

Eqs. (7.2) and (7.5) as stated are suitable for the vacuum environment and are rewritten for the material environment as follows, respectively [16]:

$$\nabla \cdot D = \rho \tag{7.7}$$

$$\nabla \times H = j + \frac{\partial D}{\partial t} \tag{7.8}$$

where H and P are the external magnetic field strength and the polarization, respectively. The amount of electrical displacement (D) is calculated by [16]:

$$D = \varepsilon_0 E + P \tag{7.9}$$

The relation between the magnetic induction and the external field in the Maxwell equations Eq. (7.5) is given by [16]:

$$B = \mu_0 (H + M) \tag{7.10}$$

In the above equation, M is the magnetization (or the magnetic field induced in the object), H is the external magnetic field strength, and B is the intensity of the field produced by the interaction of the external field and the field produced in the object.

7.4 Electromagnetic fields application in drug delivery

Drug delivery is the method or process of administering a pharmaceutical compound to achieve a therapeutic effect in humans or animals. Drug delivery system (DDS) is systems based on interdisciplinary sciences such as polymer science, pharmaceutics, bio-conjugated chemistry, and molecular biology. DDS transports pharmaceutical compounds by nanocarriers to the desired location inside the human body in a safe manner. The maximum efficiency and minimum side effect are achieved by using nanocarriers and electromagnetic fields in DDSs. The delivery of drugs into human body is designed based on several factors such as disease types, desired effect, and the product availability [1].

7.4.1   Magnetic nanocarriers for controlled drug delivery

Chemotherapy using special drugs is one of the most famous methods for treating cancer. The main task of the special drugs is to eliminate the cancer cells without damaging adjacent tissues. The goal of chemotherapy is to inject a high concentration of drug into the body. The cancer cells are more susceptible to chemotherapy drugs due to their much faster growth than the healthy cells [17].

Scientists are still studying the unique properties of cancer cells in order to increase the accuracy of targeting. During the chemotherapy process, in addition to cancer cells, other cells such as hair cells and intestinal cells, which have a high growth rate, are also attacked and their growth stops. This disrupts the patient's affairs. The most important weakness of the chemotherapy is that it is relatively nonselective toward healthy and unhealthy tissue [2]. The side effects associated with chemotherapy are the decrease in production of the blood cells, including immune system cells, painful inflammation, and ulceration of mucous membranes of the digestive system, loss of body hair, and dry skin. One way to reduce the amount of drug release in healthy tissues is to control the delivery of the drug. The controlled delivery enables the release of drug to the targeted tissue and avoid the spread of them throughout the patient's body. It also reduces the use of the required drug [2].

The macromolecular drug carriers are one of the best ways to deliver the drug because they target only the tumors and make no poisoning to the healthy tissues. One difficulty is to insert them into the cancerous tumors. The macromolecular drug carrier does not easily penetrate deep enough into the tumor and tend to integrate near the surface of the veins due to the high internal pressure of the tumor and their small penetration coefficient [18,19]. Bayern et al. provided a comparison between the results of conventional treatment and treatment in a controlled drug delivery. They showed a much more effective drug delivery in a controlled manner [20].

A method of controlling drug delivery is the use of magnetic nanocarriers along with a magnetic or an electromagnetic field. The nanocarriers is guided to the desired place by the magnetic/electromagnetic field. In this method, the drug and magnetic nanocarriers, which are usually made of iron, nickel, cobalt or their oxides, are injected into the upstream of the flow of the designated tissue [5]. The nanocarriers move along the bloodstream close to the magnetic field. Then they are accumulated near the designated tissue wall due to the inserted force by the electromagnetic field [5]. Then the drug is transmitted through the wall into the tissue. The schematic of the controlled drug delivery by magnetic field is shown in Fig. 7.4.

7.4.2   Magnetic field for controlled drug delivery

In drug delivery, the magnetic field must impose high power gradient to the targeted tissue. The magnetic field required is created by permanent or electrical magnet. The magnets that are placed outside the body adjacent to the target tissue collect the MNPs in the targeted area. The magnetic field gradient sharply decreases as it moves away from the targeted tissue and therefore it does not

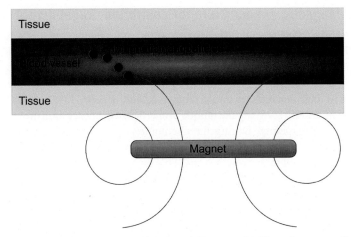

FIGURE 7.4 Simplified model of controlled drug delivery method based on magnetic nanoparticles.

penetrate depth into tissue. To overcome the penetration problem two alternative methods are utilized [9]:

Method (1) Use a few magnets to create a focal point at the desired location.

Method (2) Implant a metal near the target tissue. The metal implant is transformed into a local magnet under the influence of its external magnetic field and generates a strong magnetic field gradient.

Li et al. [9] reported higher efficiency when implant method is used compared to the use of several magnets. One drawback is that this method may cause local injury and may endanger the patient's health [21].

7.4.3 Magnetic force applied to nanocarriers

Generally, Maxwell equation is used directly to determine the electromagnetic behavior of nanocarriers and is used indirectly to obtain the force exerted on the magnetic nanocarriers in a magnetic field. It means that the amount of force applied to the particle obtained by using the gradient of the calculated magnetic energy. The magnetic force on a particle ($\vec{F}_m$) in a magnetic field is equal to the gradient of magnetic energy as follow [22]:

$$\vec{F}_m = -\nabla U_m \tag{7.11}$$

where U_m and $\vec{F}_m$ are nanocarrier magnetic energy and magnetic force vector, respectively. The magnetic energy of a single object immersed in an external magnetic field (H) is obtained from [22]:

$$U_m = -\frac{1}{2}\mu_0 \int \vec{M} \cdot \vec{H} dv \tag{7.12}$$

In above equation H, μ_0, and dv are external magnetic field, vacuum magnetic permeability, and finite element, respectively. Since the magnetic nanocarrier size is very small, the magnitude of the external magnetic field and the internal magnetic field (magnetization) of the magnetic nanocarriers are approximately uniform. As a result, the magnetic energy of a nanocarriers Eq. (7.12) is obtained from:

$$U_m = -\frac{1}{2}\mu_0 \forall_{MNP}\left(\vec{M} \cdot \vec{H}\right) \tag{7.13}$$

By combining Eqs. (7.11) and (7.13), the amount of force applied to the MNP is obtained from:

$$\vec{F}_m = \frac{1}{2}\mu_0 \forall_{MNP}\nabla\left(\vec{M} \cdot \vec{H}\right) \tag{7.14}$$

$\forall_{MNP}$ is the volume of the nanoparticle. To calculate the magnetic energy (Um), it is necessary to determine the magnitude of the internal magnetic field (magnetization) by applying Eq. (7.14). The internal magnetic field of an object is based on two inductive magnetic fields under the influence of the external magnetic field and the hysteresis. In paramagnetic materials, the induction magnetic field due to the external magnetic field is determined from the Longevin's law as follow [22]:

$$\frac{M}{M_s} = \coth\left(\frac{3\chi H}{M_s}\right) - \frac{1}{\dfrac{3\chi H}{M_s}} \tag{7.15}$$

where χ and M_s are the magnetic susceptibility and saturation magnetization, respectively. The saturation state is a situation in which the increase in the external field strength has no effect on the magnetization. The internal induction field of the nanocarriers is calculated in terms of the intensity of the external magnetic field (Eq. 7.15) and is plotted in Fig. 7.5.

As shown in Fig. 7.5, for the nonintense magnetic field ($3\chi H/M_s < 2.5$), the relation between the internal magnetic field and the strength of the external magnetic field is almost linear and hence the right-hand side of Eq. (7.15) is approximated as [22]:

$$\frac{M}{M_s} = \frac{\chi H}{M_s} \Rightarrow M = \chi H \tag{7.16}$$

Based on experiment, the magnitude of the induction magnetic field of nanocarriers is calculated from:

$$M = \frac{\chi}{1 + D_m\chi}H \tag{7.17}$$

The amount of demagnetization factor (D_m) for spherical objects is one-third. For a strong magnetic field ($H \to \infty$), saturation magnetization is obtained (M_s), and Eq. (7.15) is rewritten [22]:

$$\frac{M}{M_s} = 1 \Rightarrow M = M_s \tag{7.18}$$

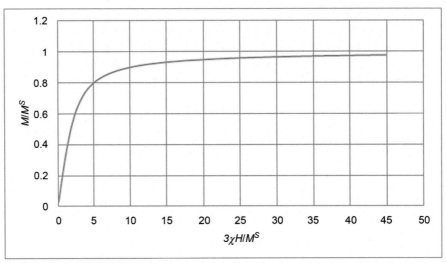

FIGURE 7.5 The internal induction field of nanocarrier in terms of the intensity of the external magnetic field.

By combining Eqs. (7.14)–(7.18), the magnetic force applied to a nanocarriers is obtained from:

$$\vec{F}_m = \frac{1}{2}\mu_0 \forall_{MNP} M_s \nabla \left[|H| \left(\coth\left(\frac{3\chi|H|}{M_s}\right) - \frac{1}{\frac{3\chi|H|}{M_s}} \right) \right] \qquad (7.19)$$

For the nonintense magnetic field ($3\chi H/M_s < 2.5$), Eq. (7.19) simplify into a useful and accurate equation as follow:

$$\vec{F}_m = \frac{1}{2}\mu_0 \forall_{MNP} \frac{\chi}{1+0.33\chi} \nabla\left(|H|^2\right) \qquad (7.20)$$

7.5   Hyperthermia

Hyperthermia therapy is one of the new types of treating cancer tissue. Hyperthermia is usually utilized in combination with other therapies such as chemotherapy or radiotherapy. In hyperthermia therapy, the cancerous tissue alone or the tissue and its surrounding are exposed to high temperature, as high as 44°C, which is higher than deep heating of the cancerous tissue, also called diathermy [23]. In this method by increasing the temperature of the cancerous tissue up to 44°C the living cancerous cells that are very sensitive to high temperature, get damaged before reproduction. Temperature higher than 45°C may completely damage the healthy and the cancerous tissue. Therefore, the healthy cell temperature during the hyperthermia treatment should stay less than 41°C [24].

7.5.1 The application of MNPs in hyperthermia therapy

The activation of MNPs by an alternating magnetic field (AMF) is currently being explored as a technique for targeted therapeutic heating of tumors. Various types of superparamagnetic and ferromagnetic particles, with different coatings and targeting agents, allow for tumor site and type specificity [25]. Alternative magnetic fields are among the methods, such as ultrasound, radio waves, microwave waves, and infrared waves that generate the needed heat for hyperthermia treatment. MNP hyperthermia is also being studied as an adjuvant to conventional chemotherapy and radiation therapy.

In this method, the targeted cancerous tissue is heated by the induced electromagnetic field. The rise in the body temperature during the MRI imaging is an example of heat produced by electromagnetic field [26].

Hyperthermia like other medical treatments have some side effect for human body. The most common limitation of conventional hyperthermia therapy is as follows [27]:

1. Increases the healthy tissue temperature as well
2. Does not raise the temperature high enough in areas with high blood flow or thick tissue coverage
3. Penetration of the wave into the deep tissue is limited

To overcome the above-mentioned limitation, some MNPs are injected into the targeted tissue during the hyperthermia therapy. The injected MNPs due to their metallic base produce more heat in external magnetic field and increase the temperature of the tissue [23]. Ho et al. investigated the effect of the MNP injections on the growth of the cancerous tumor in mice during the hyperthermia therapy [28]. They reported that the patient's recovery is much higher with nanoparticle injection compare to without nanoparticle injection.

7.5.2 Sources of heat production in MNP

MNPs produce heat when they are placed in an alternative magnetic fields. The generated heat is the sum of the three sources; eddy current heat loss, the hysteresis loss and the residual loss [29].

7.5.2.1 Eddy current heat loss

Based on electrical conductivity, alternative magnetic field produces electrical potential in MNPs. The produces electric potential creates electric vortex current at the surface of the nanoparticles and then the vortex current is converted to eddy current heat loss. The amount of the eddy current heat loss (P_e) for spherical nanoparticles is calculated by [30]:

$$P_e = \frac{\pi^2}{20} \times B_m^2 d^2 \sigma f^2 \qquad (7.21)$$

The diameter of the nanoparticle is d, the conductivity of the nanoparticle is σ, the frequency of the field is f, and the magnitude of the magnetic field induced in the material is B_m. The eddy current is negligible for nanoparticles with diameter of less than centimeter [31].

7.5.2.2 Hysteresis loss

When the ferromagnetic material is exposed to an alternating external magnetic field, its magnetization runs through a closed cycle. The closed cycle is characterized by three parameters: saturation magnetization (M_s), magnetic residual (M_r), and coercivity (H_c). The area inside the cycle represents the amount of energy given to the substance (which becomes heat) per cycle [32].

Single domain nanoparticles release the highest amount of hysteresis loss. The maximum amount of hysteresis loss per unit volume for each cycle (Q) is calculated as follows [32]:

$$Q = 4\mu_0 \cdot M_s \cdot H_c \tag{7.22}$$

The coercivity is calculated from Refs. [4,33]:

$$H_c = \frac{2K}{\mu_0 M_s}\left(1 - 5\sqrt{\frac{k_b T}{KV}}\right) \tag{7.23}$$

Anisotropic energy density is K, temperature (in Kelvin) is T, the volume of the nanoparticle is V and the Boltzmann constant is k_b. The amount of thermal energy per unit mass of nanoparticles (SLP) is obtained from [32]:

$$SLP = \frac{Q \cdot f}{\rho} \tag{7.24}$$

where ρ is nanoparticles density. The amount of heat generated in a metallic nanoparticle ($\dot{q}_p$) is determined from:

$$\dot{q}_p = f \cdot \forall \cdot Q = 4\mu_0 \cdot M_s \cdot H_c \cdot f \cdot \forall \tag{7.25}$$

The saturation magnetization and the anisotropy energy density of two samples nanoparticles, iron and oxides are given in Table 7.1 [4].

The behavior of the coercivity for different nanoparticles size is shown in Fig. 7.6.

The saturation magnetization, the coercivity, and density of MNPs, particles that are generally used in hyperthermia applications, are approximately 446, 30 kA/m and 5240 kg/m^3, respectively [32, 34]. The thermal energy of MNPs

Table 7.1 Anisotropy and crystalline parameters defining SD and SPM critical diameters at 300 K.

	M_s (kA/m)	K (kJ/m^3)
Magnetite	446	13.5
Maghemite	380	4.6

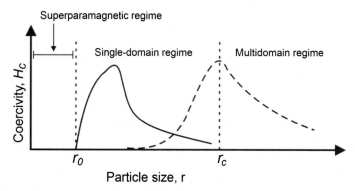

FIGURE 7.6 The coercivity changes vs the nanoparticles size.

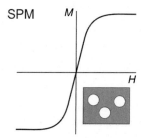

FIGURE 7.7 Induction field in superparamagnetic particles due to alternating magnetic field.

under the 500 kHz external magnetic field is calculated from Eq. (7.24) and is approximately 7 kW/g. As mentioned the residual loss has no effect on thermal energy [32].

The hysteresis loss in large single-domain nanoparticles is significant. But it sharply decreased as nanoparticle size decreases. Therefore, the hysteresis loss for small nanoparticles is negligible [31].

Fig. 7.7 depicts an induced field in superparamagnetic nanoparticles by external magnetic field. As shown superparamagnetic nanoparticles magnetic hysteresis and hence, the production of hysteresis loss in them is small [31].

7.5.2.3 Residual loss

The residual loss is produced by the relaxation processes, Neal and Brownian, in an alternative external magnetic field. The residual loss produced per unit volume is obtained from [29,32]:

$$P = \pi \mu_0 \chi'' f H^2 \tag{7.26}$$

where χ'', f, and H are the virtual magnetic susceptibility, the frequency, and the magnetic field strength, respectively. The virtual magnetic susceptibility of the material is a function of the frequency and is calculated as follow [29,32]:

$$\chi'' = \frac{2\pi f \tau_{eff}}{1+\left(2\pi f \tau_{eff}\right)^2}\chi_0 \tag{7.27}$$

The effective time constant is τ_{eff} and the magnetic susceptibility is χ_0. By inserting Eq. (7.27) in Eq. (7.26) the residual loss is obtained by [29,31,32]:

$$P = \frac{1}{2}\mu_0\chi_0 H^2 \frac{\left(2\pi f\right)^2 \tau_{eff}}{1+\left(2\pi f \tau_{eff}\right)^2} \tag{7.28}$$

As stated, the eddy current heat loss negligible and can be neglected, and the hysteresis loss and residual loss generate heat in the ferromagnetic nanoparticles. Hence, there are two ways to achieve a significant amount of heat production [32]:

1. Using a high-frequency magnetic field (several hundred kHz) and a limited amplitude (several kilo ampere/m) for superparamagnetic nanoparticles (increasing the residual loss)
2. Using a limited frequency magnetic field (several hundred kHz) and a high amplitude (several 10 kA/m) for ferromagnetic nanoparticles (increasing the hysteresis loss)

Superparamagnetic nanoparticles do not have magnetic hysteresis and hence the hysteresis loss doesn't generate. As a result, the superparamagnetic nanoparticles are the choice.

The heat in MNPs is produced due to the variable and alternative magnetic field, and for constant field no heat generates inside the nanoparticles. Also, according to the above equations, the heat loss is increased by increasing the amplitude magnetic field.

Pankhurst et al. [2] reported the allowed alternative magnetic field for therapy (Table 7.2).

Frequencies below the permissible level may stimulate bones or peripheral muscles and heart muscle. However, the higher magnetic field strength causes the accumulation of nanoparticles and embolism in vessels [24]. Rast et al. [24] and Dutz and Hergt [35] reported the same range as Pankhurst et al. [2]. Also they have proposed the product of the frequency in field strength up to the 5×10^9 A/(m s) for parts of the body with a diameter of less than 30 cm.

Table 7.2 The range of allowable values for alternating magnetic fields [2].

	Maximum	Minimum
Frequency (kHz)	1200	50
Field amplitude (kA/m)	15	–
The product of the frequency in field strength A (m s)	4.85×10^8	–

7.6   Application of external magnet on cancerous solid tumors

As mentioned magnetic field is one of the most useful external forces for moving the nano-drug carrier through the body and deliver the drug to the desired place. One of the drug carrier option is MNPs. MNPs are typically made from iron, nickel, and cobalt. They are loaded with drugs, coated with biocompatible coatings and are injected into the human vessel (bloodstream). The loaded particles move through the vessel and are absorbed by cancerous solid tumor where the magnetic field is applied. The effect of external magnet on the effectiveness of the cancerous solid tumor treatment is presented below. The physical parameters that affect the magnetic drug carriers (MDCs) distribution are studied. The MDCs diffusion coefficients in the capillary as well as its wall and the tumor tissue are considered as variable and calculated. These coefficients are functions of MDCs diameter, pore size of capillary wall, tissue porosity, etc.

Fig. 7.8 shows the schematic model of the capillary, its wall (the endothelium layer), the tumor tissue and external magnet. The external magnet is cylindrical with 4mm diameter and is placed in the middle of the space on the top of the tumor tissue.

The MNPs are core-shell shape (superparamagnetic metallic core with 5 nm biocompatible shell). The blood is treated as non-Newtonian, its density is 1050 kg/m^3 and its viscosity is calculated by the power law as follow [1]:

$$\mu = m\dot{\gamma}^{n-1} \tag{7.29}$$

where $\dot{\gamma}$ is the blood shear rate and m and n are constant and equal to 0.012 and 0.8, respectively [1].

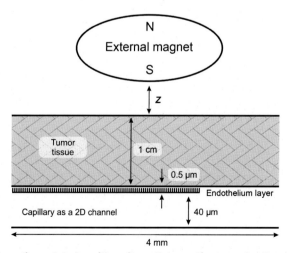

FIGURE 7.8 Schematic model of capillary, its wall, tumor tissue, and external magnet.

The continuity and momentum equations for blood are as follows:

$$\frac{\partial u}{\partial x} + \frac{\partial v}{\partial y} = 0 \tag{7.30}$$

$$\rho\left(u\frac{\partial u}{\partial x} + v\frac{\partial u}{\partial y} \right) = -\frac{\partial P}{\partial x} + \mu\left(\frac{\partial^2 u}{\partial x^2} + \frac{\partial^2 u}{\partial y^2} \right) + F_x \tag{7.31}$$

$$\rho\left(u\frac{\partial v}{\partial x} + v\frac{\partial v}{\partial y} \right) = -\frac{\partial P}{\partial y} + \mu\left(\frac{\partial^2 v}{\partial x^2} + \frac{\partial^2 v}{\partial y^2} \right) + F_y \tag{7.32}$$

The MDCs structure is depicted in Fig. 7.3. As shown the MDCs are core-shell (spherical core with 5 nm biocompatible shell). The volume of MDC ($\forall_{MDC}$) is sum of core sphere volume and coating volume. Total volume of MNPs ($\forall_{MNP}$) is 0.7 of core sphere volume. Therefore, total MNP's volume ($\forall_{MNP}$) of single MDC is as follow:

$$\forall_{MNP} = 0.7\left(\forall_{MDC} - \forall_{shell} \right) \tag{7.33}$$

The magnetic body force inside the capillary is given by [1]:

$$\vec{F} = \vec{F_1} n_{p=} 0.5 \times \forall_{MNP} \times \mu_0 \frac{\chi_{MNP}}{1 + \chi_{MNP}/3} \times \nabla | H |^2 \times n_p \quad \text{and} \quad n_p = \frac{C_{MDC}}{\forall_{MDC}} = C \times \frac{C_0}{\forall_{MDC}} \tag{7.34}$$

where $\forall_{MNP}$ is single MNP's volume and χ_{MNP} is the magnetic susceptibility of the MNPs and set equal to 3 [36]. Also C_{MDC}, C_0, and C are volumetric concentrations of MDCs in the blood, concentration of MDCs at inlet and dimensionless concentration (C_{MDC}/C_0), respectively.

The concentration equation inside the capillary is as follows [1]:

$$\frac{\partial C}{\partial t} + \nabla \cdot \left(C\vec{v}_{MDC} \right) = \nabla \cdot \left(D_{blood} \nabla C \right) \tag{7.35}$$

where D_{blood}, $\vec{v}_{MNP}$, and $V_{relative}$ are MNPs diffusion coefficient in blood, MNPs velocity vector, and the MNPs relative velocity with respect to the blood flow, respectively. Also G and E are generation and uptake terms, respectively. In this model, no generation and no uptake is considered. [1].

$$\vec{v}_{MNP} = \vec{v} + \vec{v}_{relative} \tag{7.36}$$

$$D_{Blood} = D_B + D_S \quad D_B = \frac{k_B T}{6\pi\mu_{blood} r_{MDC}} \tag{7.37}$$

$$\vec{v}_{relative} = \frac{\vec{F_1}}{6\pi\mu_{blood} r_{MDC}} = \frac{\forall_{MNP}}{12\pi\mu_{blood} r_{MDC}} \times \mu_0 \frac{\chi_{MNP}}{1 + \chi_{MNP}/3} \times \nabla | H |^2 \quad \text{in capillary}$$

where D_B is Brownian diffusion coefficient and is calculated from Einstein relation [37] and D_s is scattering diffusion coefficient and it is equal to 3.5×10^{-12} [37].

Finally, the concentration equation for blood in capillary is obtained by [37]:

$$\frac{\partial C}{\partial t} = -\nabla \cdot \left[-D_{Blood}\nabla C + C\vec{v} + C\frac{0.5\mathbb{V}_{core}}{6\pi\mu r_{MDC}}\mu_0 \frac{\chi_{MNP}}{1 + \chi_{MNP}/3}\nabla\left(|H|^2\right)\right] \qquad (7.38)$$

The last term in right hand of Eq. (7.38) is mass transfer due to magnetic force (influence of external magnet). As shown this term is function of $\left(\dfrac{\mathbb{V}_{MNP}}{r_{MDC}}\right)$. By inserting $\mathbb{V}_{MNP}$ from Eq. (7.33) the term $\left(\dfrac{\mathbb{V}_{MNP}}{r_{MDC}}\right)$ is as follows:

$$\frac{\mathbb{V}_{MNP}}{r_{MDC}} = \frac{0.7\left(\mathbb{V}_{MDC} - \mathbb{V}_{shell}\right)}{r_{MDC}} = \frac{2.8\,\pi\left(r_{MDC} - 5e(-9)\right)^3}{3\,r_{MDC}} \qquad (7.39)$$

This means that this term becomes greater as MDCs radius (r_{MDC}) *increases*.

Capillary wall (endothelium) is modeled as a porous media. The concentration equation as follows [37]:

$$\frac{\partial C}{\partial t} = -\nabla \cdot \left[-D_{Endo}\nabla C + C\vec{v}_{MDC}\right] + \frac{G}{\varepsilon} - E(C) \qquad (7.40)$$

Also, it is assumed that the fluid is motionless in endothelium layer and the tissue where the maximum velocity is very small (0.016 μm/s for 1 cm radius tumor surrounded with normal tissue) [37].

Endothelium diffusion coefficient (D_{Endo}) is given by [1]:

$$D_{endo} = D_\infty \times \left(\frac{\varepsilon}{\lambda_g^2}\right) \times S \times J \qquad (7.41)$$

where D_∞ is diffusion coefficient of particle in unbounded fluid.

The gaps between endothelial cells are filled with plasma. The diffusion coefficient of particle in unbounded fluid, D_∞, is given by Brownian diffusion coefficient of particles in the plasma as follow:

$$D_{plasma} = \frac{k_B T}{6\pi\mu_{plasma}r_{MDC}} \qquad (7.42)$$

μ_{plasma} is plasma viscosity and it is equal to1.24 mPa s [38].

Also S (steric coefficient), J (hydrodynamic coefficient) and ε (the porosity of endothelium layer) are calculated as follow [37]:

$$S = (1 - \alpha)^2 \quad \alpha = \frac{r_{MNP}}{r_{core}} \qquad (7.43)$$

$$J = \left(1 - 2.1044\alpha + 2.089\alpha^3 - 0.948\alpha^5\right) \qquad (7.44)$$

$$\varepsilon = \frac{\text{Intercellular gap in Endotheium layer}}{\text{Average size of Endotheium cell}} \qquad (7.45)$$

The concentration equation for tumor tissue, a porous media, is:

$$\frac{\partial C}{\partial t} = -\nabla \times \left[-D_{Tissue}\nabla C + C\;\varepsilon\left(\frac{1}{\lambda_g^2}\right) s\;\times J\frac{\mathbb{V}_{MNPe}}{12\pi\mu_{plasma}r_{MDC}}\mu_0 \frac{\chi_{MNP}}{1 + \chi_{MNP}/3}\nabla\left(|H|^2\right)\right] \qquad (7.46)$$

D_{Tissue} is the MNPs diffusion coefficient in the tissue and is obtained by:

$$D_{Tissue} = D_\infty \times \left(\frac{\varepsilon}{\lambda_g^2}\right) \times S \times J \tag{7.47}$$

Because the interstitial fluid is motionless the D_∞ is usually equal to D_{plasma} [1]. The steric coefficient (S) and hydrodynamic coefficient (J) are obtained from [37]:

$$S = \exp\left(-0.84k^{1.09}\right), \quad k = \left(1 + \frac{r_{MDC}}{r_{fiber}}\right)^2 \times \phi \tag{7.48}$$

$$J = e^{-\alpha \times \phi^a} \tag{7.49}$$

Φ is the fibers volume fraction. Collagen fibrils radius, $r_{fibrils}$, is equal to 50 nm. The value of $r_{fibrils}$ is between 15 and 100 nm [37]. Also, the value of the fiber volume fraction of fibers, Φ is equal to 0.01 where the values for three different tumors are $\Phi = 0.01$ (LS174T), $\Phi = 0.03$ (HSTS26T), and $\Phi = 0.045$ (U87) [39].

The geometrical tortuosity (λ_g) is expressed as a function of porosity [1]:

$$\lambda_g = \varepsilon^{-n} \tag{7.50}$$

Tissue porosity, ε, varies between 0.06 and 0.6 for different tumors [1]. The value of n has an upper and lower limit and is determined by:

$$\text{Upper limit}: n = 0.23 + 0.3\varepsilon + \varepsilon^2 \quad \text{and} \quad \text{Lower limit}: n = 0.23 + \varepsilon^2 \tag{7.51}$$

For the safe simplicity an average value of n is utilized.

For initial condition dimensionless MDCs concentration (C) inside the vessel is equal to 1 and no concentration is assumed in the tumor tissue and in the capillary wall ($C = 0$).

According to Fig. 7.8 the boundary conditions are as provided in Table 7.3.

Fig. 7.9 shows the distribution of dimensionless MDCs concentration in the tumor tissue at 24 h time under influence of external magnet with different flux densities. MDCs diameters are 50, 100, 166, 250 and 345 nm.

As shown, in the absence of external magnet (0 T), the small MDCs penetrate deeper into the tumor, but their penetration into the tumor tissue, in general, is not

Table 7.3 Fig. 7.8 boundary conditions.

Boundary	Conditions
Channel inlet	Steady velocity ($U_{in} = 0.2$ mm/s [35]) Steady MDCs inlet concentration ($C_0 = 10^{-4}$)
Channel upper wall	No-slip condition ($u,v = 0$) MNPs can go through the wall
Channel lower wall	No-slip condition ($u,v = 0$) MNPs can't go through the wall
Channel outlet	Neumann boundary condition for both momentum and concentration equations
Tissue upper, right and left wall	MDCs cannot go through the wall

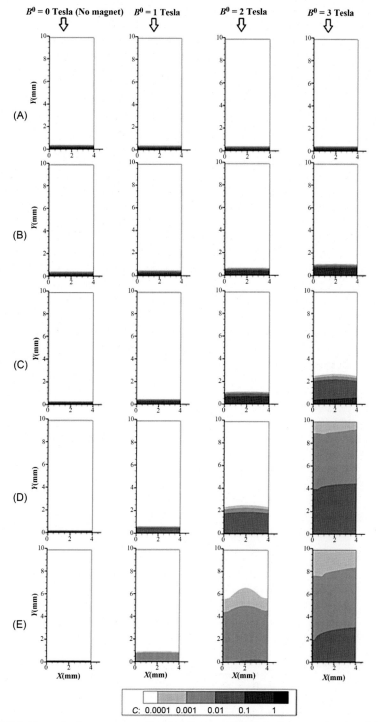

FIGURE 7.9 Dimensionless concentration distribution of MDCs in the tumor tissue at 24 h time under influence of external magnet with different flux density (A) 50 nm, (B) 100 nm, (C) 166 nm, (D) 250 nm, (E) 345 nm MDCs (porosity = 0.3, z = 5 mm).

significant and limited to tumor surface (less than 3.2% of tumor thickness, about 0.314 mm for 50 nm MDCs). This is because in absence of external magnet the diffusion is single penetration mechanism and also MDCs diffusion coefficient in endothelium layer and tumor tissue is low and also decreases as MDCs become greater. Furthermore, for MDCs greater than 50 nm, the penetration depth increases exponentially as the external magnet flux density (B) increases. However, the external magnet flux density increases from 0 to 3 T, penetration depth of 250 and 345 nm MDCs multiplied about 56 and 90, respectively. This means that the magnetic penetration becomes dominant penetration mechanism as external magnet becomes strong. This is due to the magnetic field intensity (H), and so magnetic force acting upon the MDCs increases as the external magnet flux density (B_0) increases.

Also it is observed that the penetration depth of 345 nm MDCs is about 10 times of 50 nm MDCs while external magnet flux density is 2 T. This means that the greater MDCs are more penetrative while strong external magnet is applied. This is because magnetic force acting on the MDCs exponentially increases as MDCs radius increases. Also the DMCs reached into the whole tumor as well as the concentration of great MDCs is high. This result provides the better drug delivery with minimum side effects [paper].

Figs. 7.10 and 7.11 represent the time variation of MDCs penetration depth without and with external magnet, respectively.

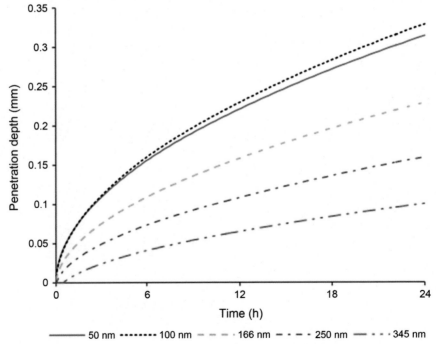

FIGURE 7.10 MDCs penetration depth in the absence of external magnet (porosity = 0.3, $z = 5$ mm).

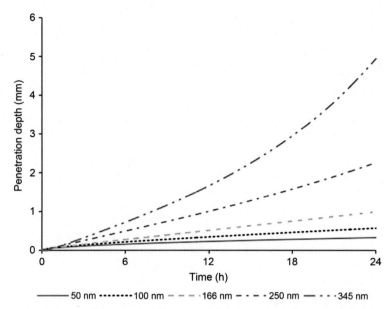

FIGURE 7.11 MDCs penetration depth in the presence of 2 T external magnet (porosity = 0.3, $z = 5$ mm).

As shown, in the absence of external magnet, the rate of the penetration depth decreases excessively as time passes. This is because the gradient of MDCs concentration which causes the diffusion decreases as time passes. However, the presence of external magnet causes the penetration depth of MDCs to increase constantly. This is because the magnetic term of MDCs penetration does not decrease with time.

References

[1] M. Ghassemi, A. Shahidian, Nano and Bio Heat Transfer and Fluid Flow, Academic Press, (2017).

[2] Q.A. Pankhurst, J. Connolly, S.K. Jones, J. Dobson, Applications of magnetic nanoparticles in biomedicine, J. Phys. D: Appl. Phys. 36 (13) (2003) R167.

[3] P.J. Cregg, K. Murphy, A. Mardinoglu, Calculation of nanoparticle capture efficiency in magnetic drug targeting, J. Magn. Magn. Mater. 320 (23) (2008) 3272–3275.

[4] D.K. Rajan, J. Lekkala, Coercivity Weighted Langevin Magnetisation; A New Approach to Interpret Superparamagnetic and Nonsuperparamagnetic Behaviour in Single Domain Magnetic Nanoparticles, 2013, arXiv preprint arXiv:1308.2517.

[5] M.R. Habibi, Numerical Study of Magnetic Nano-Particles in Blood Flow Under Non-Uniform Magnetic Field (Ph.D. Dissertation), K.N. Toosi Universiyu of Technology, Iran, 2012.

[6] A. Weddemann, I. Ennen, A. Regtmeier, C. Albon, A. Wolff, K. Eckstädt, et al. Review and outlook: from single nanoparticles to self-assembled monolayers and granular GMR sensors, Beilstein J. Nanotechnol. 1 (1) (2010) 75–93.

[7] C. Sun, J.S. Lee, M. Zhang, Magnetic nanoparticles in MR imaging and drug delivery, Adv. Drug Deliv. Rev. 60 (11) (2008) 1252–1265.

[8] J. Chomoucka, J. Drbohlavova, D. Huska, V. Adam, R. Kizek, J. Hubalek, Magnetic nanoparticles and targeted drug delivering, Pharmacol. Res. 62 (2) (2010) 144–149.

[9] R. Fernández-Pacheco, J.G. Valdivia, M.R. Ibarra, Magnetic nanoparticles for local drug delivery using magnetic implants, Micro and Nano Technologies in Bioanalysis, Humana Press, Totowa, NJ, 2009, pp. 559–569.

[10] A. Nacev, C. Beni, O. Bruno, B. Shapiro, The behaviors of ferromagnetic nano-particles in and around blood vessels under applied magnetic fields, J. Magn. Magn. Mater. 323 (6) (2011) 651–668.

[11] L.F. Gamarra, W.M. Pontuschka, E. Amaro Jr., A.J.D. Costa-Filho, G.E.D.S. Brito, E.D. Vieira, et al. Kinetics of elimination and distribution in blood and liver of biocompatible ferrofluids based on Fe_3O_4 nanoparticles: an EPR and XRF study, Mater. Sci. Eng. C 28 (4) (2008) 519–525.

[12] M.A. Dobrovolskaia, A.K. Patri, J. Simak, J.B. Hall, J. Semberova, S.H. De Paoli Lacerda, S.E. McNeil, Nanoparticle size and surface charge determine effects of PAMAM dendrimers on human platelets in vitro, Mol. Pharm. 9 (3) (2011) 382–393.

[13] B. Chertok, B.A. Moffat, A.E. David, F. Yu, C. Bergemann, B.D. Ross, V.C. Yang, Iron oxide nanoparticles as a drug delivery vehicle for MRI monitored magnetic targeting of brain tumors, Biomaterials 29 (4) (2008) 487–496.

[14] CICLO, XXIII, Biomedical Applications of Electromagnetic Fields: Human Exposure, Hyperthermia and Cellular Stimulation, Universita' Di Padova Facolta' Di Ingegneria, Dipartimento di Ingegneria dell'Informazione, Scuola di Dottorato di Ricerca in Ingegneria dell'Informazione Indirizzo: Bioingegneria, 2012.

[15] R. Fitzpatrick, Maxwell's Equations and the Principles of Electromagnetism, Jones & Bartlett Publishers, (2008).

[16] R. Fitzpatrick, Maxwell's Equations and the Principles of Electromagnetism, Jones & Bartlett Publishers, (2008).

[17] C.S. Brazel, Magnetothermally-responsive nanomaterials: combining magnetic nanostructures and thermally-sensitive polymers for triggered drug release, Pharm. Res. 26 (3) (2009) 644–656.

[18] M.R. Dreher, W. Liu, C.R. Michelich, M.W. Dewhirst, F. Yuan, A. Chilkoti, Tumor vascular permeability, accumulation, and penetration of macromolecular drug carriers, J. Natl. Cancer Inst. 98 (5) (2006) 335–344.

[19] T. Stylianopoulos, M.Z. Poh, N. Insin, M.G. Bawendi, D. Fukumura, L.L. Munn, R.K. Jain, Diffusion of particles in the extracellular matrix: the effect of repulsive electrostatic interactions, Biophys. J. 99 (5) (2010) 1342–1349.

[20] J.D. Byrne, T. Betancourt, L. Brannon-Peppas, Active targeting schemes for nanoparticle systems in cancer therapeutics, Adv. Drug Deliv. Rev. 60 (15) (2008) 1615–1626.

[21] Q. Cao, X. Han, L. Li, Enhancement of the efficiency of magnetic targeting for drug delivery: development and evaluation of magnet system, J. Magn. Magn. Mater. 323 (15) (2011) 1919–1924.

[22] J. Berthier, P. Silberzan, Microfluidics for Biotechnology, Artech House, (2006).

[23] Q. Wang, Z.S. Deng, J. Liu, Theoretical evaluations of magnetic nanoparticle-enhanced heating on tumor embedded with large blood vessels during hyperthermia, J. Nanopart. Res. 14 (7) (2012) 974.

[24] L. Rast, J.G. Harrison, Computational modeling of electromagnetically induced heating of magnetic nanoparticle materials for hyperthermic cancer treatment, PIERS Online 6 (7) (2010) 690–694.

[25] A.J. Giustini, A.A. Petryk, S.M. Cassim, J.A. Tate, I. Baker, P.J. Hoopes, Magnetic nanoparticle hyperthermia in cancer treatment, Nano Life 1 (01n02) (2010) 17–32.

[26] S. Kikuchi, K. Saito, M. Takahashi, K. Ito, Temperature elevation in the fetus from electromagnetic exposure during magnetic resonance imaging, Phys. Med. Biol. 55 (8) (2010) 2411.

[27] C.S. Kumar, F. Mohammad, Magnetic nanomaterials for hyperthermia-based therapy and controlled drug delivery, Adv. Drug Deliv. Rev. 63 (9) (2011) 789–808.

[28] C.H. Hou, S.M. Hou, Y.S. Hsueh, J. Lin, H.C. Wu, F.H. Lin, The in vivo performance of biomagnetic hydroxyapatite nanoparticles in cancer hyperthermia therapy, Biomaterials 30 (23–24) (2009) 3956–3960.

[29] Y. Zhang, Y. Zhai, Magnetic induction heating of nano-sized ferrite particle, in: Advances in Induction and Microwave Heating of Mineral and Organic Materials, IntechOpen, 2011.

[30] J. Smith, H.P.J. Wijn, Ferrites, Cleaver-Hume Press Ltd, London, (1959).

[31] A.E. Deatsch, B.A. Evans, Heating efficiency in magnetic nanoparticle hyperthermia, J. Magn. Magn. Mater. 354 (2014) 163–172.

[32] S. Dutz, R. Hergt, Magnetic nanoparticle heating and heat transfer on a microscale: basic principles, realities and physical limitations of hyperthermia for tumour therapy, Int. J. Hyperther. 29 (8) (2013) 790–800.

[33] Y.W. Jun, J.W. Seo, J. Cheon, Nanoscaling laws of magnetic nanoparticles and their applicabilities in biomedical sciences, Acc. Chem. Res. 41 (2) (2008) 179–189.

[34] R.E. Rosensweig, Heating magnetic fluid with alternating magnetic field, J. Magn. Magn. Mater. 252 (2002) 370–374.

[35] R. Hergt, S. Dutz, Magnetic particle hyperthermia—biophysical limitations of a visionary tumour therapy, J. Magn. Magn. Mater. 311 (1) (2007) 187–192.

[36] T. Lunnoo, T. Puangmali, Capture efficiency of biocompatible magnetic nanoparticles in arterial flow: a computer simulation for magnetic drug targeting, Nanoscale Res. Lett. 10 (1) (2015) 426.

[37] S.M.A. Ne'mati, M. Ghassemi, A. Shahidian, Numerical investigation of drug delivery to cancerous solid tumors by magnetic nanoparticles using external magnet, Transport Porous Med. 119 (2) (2017) 461–480.

[38] U. Windberger, A. Bartholovitsch, R. Plasenzotti, K.J. Korak, G. Heinze, Whole blood viscosity, plasma viscosity and erythrocyte aggregation in nine mammalian species: reference values and comparison of data, Exp. Physiol. 88 (3) (2003) 431–440.

[39] S. Ramanujan, A. Pluen, T.D. McKee, E.B. Brown, Y. Boucher, R.K. Jain, Diffusion and convection in collagen gels: implications for transport in the tumor interstitium, Biophys. J. 83 (3) (2002) 1650–1660.

Ultrasound applications in cancer therapy

8

Chapter outline

8.1 Ultrasound in biomedical engineering

Ultrasound is a form of mechanical energy. Mechanical vibration at increasing frequencies is known as sound energy. The normal human sound range is from 16 Hz to something approaching 15–20,000 Hz. Sound waves are longitudinal waves consisting of areas of compression and rarefaction. As the energy within the sound wave is passed to the material, it will cause oscillation of the particles of that material. The mechanical vibration sounds with frequency higher than the level of human hearing is ultrasound. The frequencies used in therapy are typically between 1.0 and 3.0 MHz. Three important factors in ultrasound are frequency, wavelength, and velocity. Ultrasound wave frequency (f) is typically 1 or 3 MHz. In an "average tissue" the wavelength (l) would be 1.5 and 0.5 mm at 1 and 3 MHz, respectively. In a saline solution, the velocity of US (v) is approximately 1500 m/s compared with approximately 350 m/s in air. The velocity of US in most tissues is thought to be similar to that in saline. The mathematical representation of the relationship is $V = F \cdot \lambda$

Each material, as well as, tissue has a specific impedance to the passage of sound waves. The specific impedance of a tissue will be determined by its density and elasticity. Clearly, in the case of US passing from the generator to the tissues and then through the different tissue types, this cannot actually be achieved. The greater the difference in impedance at a boundary, the greater the reflection that will occur, and therefore, the smaller the amount of energy that will be transferred. To minimize this difference, a suitable coupling medium has to be utilized. If even a small air gap exists between the transducer and the skin the proportion of US which will be reflected approaches 99.998% which in effect means that there will be no transmission. The coupling media used in this context include water, various oils, creams, and gels. Ideally, the coupling medium should be fluid so as to fill all available spaces, and should allow transmission of US with minimal absorption, attenuation, or disturbance. At the present time, the gel-based media appear to be preferable to the oils and creams. Also, water is a good media and can be used as an alternative but clearly, it fails to meet the above criteria in terms of its viscosity. In addition to the reflection that occurs at a boundary due to differences in impedance, there will also be some refraction if the wave does not strike the boundary surface at 90 degrees. Essentially, the direction of the US beam through the second medium will not be the same as its path through the original medium. The critical angle for US at the skin interface appears to be about 15 degrees.

The absorption of US energy follows an exponential pattern and more energy is absorbed in the superficial tissues than in the deep tissues. In order to achieve certain effect, the US dosages must be considered at some point. The penetration of ultrasound waves ranges from kHz to MHz frequency levels, depending on the type of tissue and plan of treatment for the disease. By increasing the US beam penetration into the tissues, the less energy is available to achieve therapeutic effects. The half-value depth will be different for each tissue and also for different US frequencies. Table 8.1 gives some indication of average half value depths for therapeutic ultrasound [1].

It is not possible to know the thickness of each tissue layer in an individual patient. To achieve a particular US intensity at depth, account must be taken of the proportion of energy which has been absorbed by the tissues in the more superficial layers. For example, the energy level at 3 and 1 MHz frequencies is decreased about 25% in depth of 4 and 8 cm of typical tissue respectively. Ultrasound frequencies are used for medical application is summarized in Fig. 8.1.

The penetration (or transmission) of US is not the same in each tissue type. Generally, the tissues with the higher protein content will absorb US to a greater extent, thus tissues with high water content and low protein content such as blood and fat, absorb little of the US energy while the tissue with a lower water content and

Table 8.1 Indication of average half value depths for therapeutic ultrasound.

Frequency	Fat	Muscle	Tendon
1 MHz	50 mm	9 mm	6.2 mm
3 MHz	16.5 mm	3 mm	2 mm

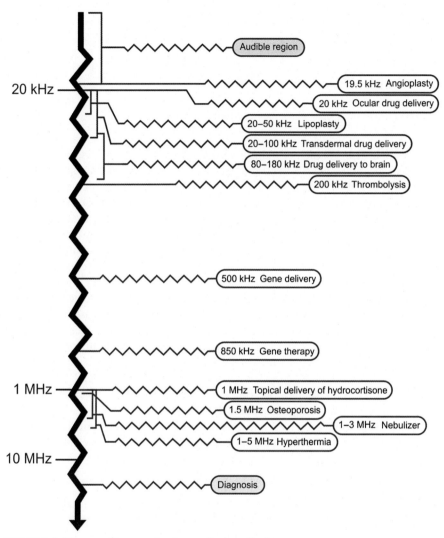

FIGURE 8.1 Ultrasound frequencies for medical application.

a higher protein content will absorb US far more efficiently [2]. Fig. 8.2 depicts the ultrasound absorption in different tissues.

Bone is at the upper end of this scale, due to the majority of US energy is reflected from them. The best absorbing tissues are those with high collagen content include ligament, scar tissue, fascia, tendon, and joint capsule [3]. The application of therapeutic US to tissues with a more highly absorbing material is more effective than the other tissues.

Ultrasonic waves have many applications in the field of health. The most important applications are the diagnosis and treatment of the disease, drug delivery, and

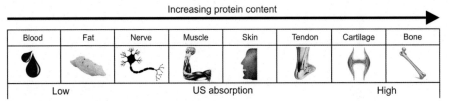

FIGURE 8.2 The ultrasound absorption in different tissues.

cell division. Transmission of ultrasound waves from tissues and organs can lead to chemical reactions, physical and biological changes. Today, many applications of ultrasonic waves have been proposed. The use of acoustophoresis technology for applications such as cell separation [4], cell trapping [5], blood plasma separation, forensic analysis [6], food analysis, cell sorting and cell synchronization [7], cell differentiation, and cellular compression. Other sections of the study have been dedicated to diagnosis of diseases and the study of organs by ultrasound imaging. In the ultrasound imaging, probability of pregnancy, size and sex of the fetus, the appearance of tumors and cysts, the internal structure of the heart, kidney stones, prostate cancer, thyroid gland, and fluid accumulation in the lung are examined. Increasing blood flow, muscle cramps decrease, fibroblast stimulation, increased protein production, increased tissue regeneration, and bone repair are some of therapeutic applications of ultrasonic waves. Ultrasonic surgeries are mainly used for cutting soft tissues (by using the thermal effect) and destroying hard tissues, by centralized ultrasonic force. Other ultrasonic applications are; drug delivery, gene and DNA transfer [8]. The working frequencies, the power, and pressure of the ultrasound waves in medical applications are shown in Table 8.2 [9].

Medical ultrasound divided into two categories: diagnostic and therapeutic. A noninvasive diagnostic technique used to image inside the body is called diagnostic

Table 8.2   Different frequencies in medical applications of ultrasound.

Application	Frequency (MHz)	Moderate intensity (W/cm^2)	Maximum Pressure (MPa)
Sonography mode B	1–15	Less than 1	0.45–5.5
Sonography mode M	3–10	0.3	More than 4
Echo	3–7.5	1	2.5
Physiotherapy	1	1	Less than 0.5
Lithotripsy	0.5–1	Few	More than 20
cutting soft tissues	0.25	Few	5–30
Highly focused ultrasound waves	0.5–5	1000–10,000	10
Bone repair	1.5	0.03	0.05
Drug delivery	2	Variable	0.2–8

ultrasound. The ultrasound probes (transducers) almost are placed on the skin (probes may be placed inside the body) and operate at megahertz (MHz) range frequencies.

The functional and anatomical ultrasound are two categories of diagnostic ultrasound. The anatomical ultrasound generates images of internal organs or other structures, but the functional ultrasound creates information maps by combining data about the movement and velocity of tissue or blood, softness or hardness of tissue, and other physical characteristics, with anatomical images. Another functional form of ultrasound is elastography, a method for measuring and displaying the relative stiffness of tissues, which can be used to differentiate tumors from healthy tissue. Elastography can be used to test for liver fibrosis, a condition in which excessive scar tissue builds up in the liver due to inflammation. Diagnostic ultrasound is generally regarded as safe and does not produce ionizing radiation like that produced by X-rays. Still, ultrasound is capable of producing some biological effects in the body under specific settings and conditions.

The purpose of the *therapeutic ultrasound*, as a noninvasive method, is to interact with tissues in the body such that they are either modified or destroyed without produces images. Some modifications such as moving or pushing tissue, heating tissue, dissolving blood clots, or delivering drugs to specific locations in the body are made possible by the use of very high-intensity beams that can destroy diseased or abnormal tissues such as tumors.

Therapeutic ultrasound produces high levels of acoustic output that can be focused on specific targets for the purpose of heating, ablating, or breaking up tissue. One type of therapeutic ultrasound uses high-intensity beams of sound that are highly targeted and is called high intensity focused ultrasound (HIFU). HIFU is being investigated as a method for modifying or destroying diseased or abnormal tissues inside the body without having to open or tear the skin or cause damage to the surrounding tissue.

Using ultrasound waves in the body can lead to heat, bubbles, tension, and vibration. In the same intensity, the heat, tension, and vibration increases with increasing frequency of waves, while the possibility of bubble formation decreases. Therefore, determining the appropriate frequency for medical application of ultrasound waves requires careful examination. These four essential effects can be beneficial or harmful. In the lithotripsy, cancer treatment and drug delivery to brain, the tension and vibration, generated heat and formation of bubbles are important, respectively [10, 11].

As the pressure amplitude, the frequency, or the propagation length is increased, the ultrasound wave could ultimately lead to a discontinuity or shock in the waveform, increasing frequency, nonlinear acoustic distortion, or pulse length can increase heating and enhance some nonthermal mechanisms, for example, radiation force. But decreasing frequency increases the chance of cavitation. Also Increasing power or intensity tends to increase the magnitude of the bioeffect mechanisms. Therapeutic ultrasound devices use continuous waves to deliver effective ultrasonic energy to tissues. Some devices operate at higher amplitude and therefore, tend to produce shocked or distorted waves. Ultrasound-induced heating is the result of the absorption of ultrasonic energy in biological tissue. For diagnostic ultrasound, temperature raises are kept relatively low or negligible by applying the limited temporal average intensities, and generally short exposure durations. Therapeutic applications of ultrasonic heating, utilize longer durations of heating with unfocused beams, or utilize higher intensity (than diagnostic) focused ultrasound. The physical therapy is a good example for the use of unfocused heating, to produce enhanced healing without injury in bone or tendon. Depending on the temperature gradients in ultrasound therapy, the effects of mild heating, coagulative necrosis, tissue vaporization, or all three are reported.

Ultrasonic cavitation and gas body activation are closely related mechanisms which depend on the rarefactional pressure amplitude (about several MPa) of ultrasound waves. This tensile stress is supported by the medium and, for example, a 2-MPa rarefactional pressure, which is common even for diagnostic ultrasound. This high rarefactional pressure can act to initiate cavitation activity in tissue when suitable cavitation nuclei are present, or directly induce pulsation of preexisting gas bodies, such as occur in lung, intestine, or with ultrasound contrast agents.

For high-power or high-amplitude ultrasound for therapy, several different mechanisms may be contributing concurrently to the total biological impact of the treatment. In addition to direct physical mechanisms for bioeffects, there are secondary effect such as vasoconstriction, extravasation, ischemia, and immune responses [12], that are greater than the direct insult from the ultrasound.

8.2 Approved modes for ultrasound therapy

Ultrasound has been used for therapy since the 1930s. Early applications were applied various mechanisms and conditions for tissue heating [13]. The use of therapeutic ultrasound was established for physiotherapy, applications in neurosurgery and for cancer treatment, in the 1970s, 1977, and 1979, respectively [12]. The application of ultrasound for therapeutic efficacy also carries the risk of unintentional adverse bioeffects so standardization, ultrasound dosimetry, benefits assurance, and side-effects risk minimization are necessary.

Therapeutic applications of ultrasound may be used clinically after government approval for marketing suitable treatment devices. A list of therapy applications with FDA approved devices in clinical use is provided in Table 8.3. The therapeutic effects of US are generally divided into thermal which include physical therapy, hyperthermia and high-intensity focused ultrasound and nonthermal including extracorporeal shock wave lithotripsy, intracorporeal lithotripsy, and lower power kilohertz frequency ultrasound devices. Also new methods of therapeutic ultrasound are including new microbubble- or cavitation-based treatment methods.

8.2.1 Thermal ultrasound therapeutic applications

The important thermal ultrasound therapeutic applications are included as physical therapy, hyperthermia, and HIFU.

Table 8.3　A list of therapy applications with FDA approved.

Therapy method	Frequency	Therapeutic outcome	Bioeffect mechanism
Unfocused beam	1–3 MHz	Tissue warming	Heating
Hyperthermia	1–3.4 MHz	Cancer therapy	Regional heating
HIFU	0.5–2 MHz	Uterine fibroid ablation	Thermal lesion
	4 MHz	Laparoscopic tissue ablation	Thermal lesion
	3.8–6.4 MHz	Laparoscopic or open surgery	Thermal lesion
	4.6 MHz	Glaucoma relief	Permeabilization
Focused ultrasound	4.4–7.5 MHz	Skin tissue tightening	Thermal lesion
Extracorporeal lithotripsy	~150 kHz	Kidney stone comminution	Mechanical stress, cavitation
Intracorporeal lithotripsy	25 kHz	Kidney stone comminution	Mechanical stress, cavitation
Extracorporeal shock wave	~150 kHz	Plantar fasciitis, epicondylitis	Unknown
Phacoemulsification	40 kHz	Lens removal	Vibration, cavitation
US assisted liposuction	20–30 kHz	Adipose tissue removal	Fat liquefaction, cavitation
Tissue cutting and vessel sealing	55.5 kHz	Laparoscopic or open surgery	Thermal lesion, vibration
Intravascular US	2.2 MHz	Thrombus dissolution	Unknown, gas body activation
Skin permeabilization	55 kHz	Transdermal drug delivery	Unknown
Low-intensity pulsed US	1.5 MHz	Bone fracture healing	Unknown

8.2.1.1　Physical therapy

Unfocused beams of ultrasound for physical therapy were the first clinical application, dating to the 1950s, which often has been referred to simply as "therapeutic ultrasound" [14]. This modality now typically has a base unit for generating an electrical signal and a hand-held transducer. The hand-held transducer is applied with coupling gel and moved in a circular motion over an injured or painful area of the anatomy to treat conditions such as bursitis of the shoulder or tendonitis, by trained physical therapy technicians. The objective is to warm tendons, muscle and other tissue to improve blood flow and accelerate healing. Ultrasound application can also assist by promoting transport of the compound into the skin, a method sometimes called sonophoresis or phonophoresis (as opposed to electrophoresis) [15]. Drugs such as lidocaine or cortisol have been used extensively in sports medicine. The level of clinical benefit to the patient from physical therapy ultrasound treatments remains uncertain [14, 16]. However, the risk of harm such as burns appears to be low when the modality is properly applied. Overall, ultrasound for physical therapy has therefore provided a modest level of efficacy and patient benefit, but also a low level of risk.

8.2.1.2 Hyperthermia

A new method by using ultrasonic waves to heat relatively large volumes of tissue for the purpose of cancer therapy is developed science 1980s and 1990s. This method of hyperthermia involves uniformly heating a tumor to about 42°C for periods of about 1 h, which appears to be effective in reducing tumor growth [17]. In some cases of hyperthermia therapy, the cancerous tissue alone or the tissue and its surrounding is exposed to high temperature, as high as 44°C and the healthy cell temperature of the body during the hyperthermia treatment should stay less than 41°C. In clinical trials, hyperthermia was used with or without radiation therapy and modest efficacy has been reported [18]. Research suggests that hyperthermia may be advantageous for drug delivery treatment using nanoparticles. However, the HIFU is applied to hyperthermia cancer treatment instead of the moderate-temperature hyperthermia in clinical usage.

8.2.1.3 High intensity focused ultrasound

One of the noninvasive methods of using ultrasound for the treatment of cancer that has been considered in recent years, is HIFU, that uses nonionizing ultrasonic waves to heat tissue. HIFU can be used to increase the flow of blood or lymph, or to destroy tissue, such as tumors, through a various mechanisms. By increasing the temperature in a short time and in focal zone, destruction of cancer cells without damage to healthy tissues is possible. This method was initially studied clinically for thermal ablation of inoperable brain tissue for Parkinson's disease [19]. The continuous lower frequencies are used to achieve the necessary thermal doses. However, pulsed waves may also be used if mechanical (such as lithotripsy) rather than thermal damage is desired. The necessary intensity at the target tissue without damaging the surrounding tissue is achieved by using the acoustic lenses. HIFU may be combined with other imaging techniques such as medical ultrasound or magnetic resonance imaging (MRI) to enable guidance of the treatment and monitoring.

In expressing the major advantage of using ultrasound waves in treating cancer compared to other therapies, can mention as follows [20]:

- Noninvasive
- Increase temperature only in the focal area and reduce the chance of damage to adjacent tissues and skin
- Nonionizing tissue and no limitation in number of radiation
- Can be used in treating most of the cancerous tissues

In a HIFU system, a signal generator is connected to a focusing transducer, which produces very high local intensities of >1 kW/cm^2 of 0.5–7 MHz ultrasound at the focal spot (see Fig. 8.3). HIFU application in therapy and treatment of disease is one

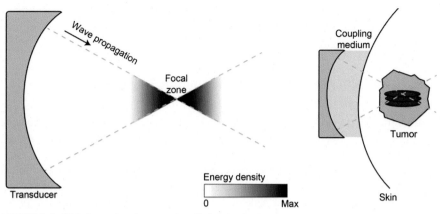

FIGURE 8.3 High-intensity ultrasound emission in tissue.

of the more active areas of research and development among all the nonionizing-energy modalities such as radiofrequency, lasers, and microwaves. One of the oldest and possibly the most investigated area of HIFU, is the treatment of benign prostatic hyperplasia and the treatment of prostate cancer. A number of research has established the use of HIFU as available option for the management of prostate cancer [21, 22]. The capability to focus energy several millimeters to centimeters away from the transducer plane is available in HIFU system. Therefore, the location of the treatment zone is accurately determined. The focused ultrasound beam can then be moved to a different location to complete the treatment of the planned volume. MRI and ultrasound imaging are used for image guidance and treatment monitoring. The uterine fibroid treatment, breast cancer, and prostate cancer management have ultrasound therapy sub-systems integrated into MR-imagers [12].

The ultrasound image monitoring of tissue changes during ultrasound therapy is based on a combination of speed of sound, attenuation, stiffness, and vapor content changes in the target region [23], including boiling detection and combined measurement and modeling approaches [12]. In addition to external focused devices, a number of other devices and systems are being developed for soft tissue coagulation which are primarily used in noninvasive approaches, or through natural orifices such as the transrectal approach for prostate treatments [24]. For example, transurethral ultrasound has been proposed for heating the prostate [25], and endoscopic treatment using an intraductal ultrasound probe has been used to treat bile duct tumors. Treatment of hepatic and pancreatic cancer can also lead to serious complications, including fistula formation and rib necrosis with delayed rib fracture [26]. Depending on the device, as well as the cosmetic application, both thermal, as well as nonthermal mechanisms within an ultrasound field are employed for these procedures. Like other methods, HIFU applications have bioeffect complications can also occur with unique risk-benefit considerations for each application, such as unwanted burns and pain, cause vasospasm and hemorrhage under conditions which generate concomitant cavitation in tissue. Detailed safety considerations should accompany the introduction of HIFU applications into clinical practice in order to assure benefit while minimizing risk to the patient.

8.2.2   Non-thermal ultrasound therapeutic applications

8.2.2.1   Extracorporeal shock wave lithotripsy

Extracorporeal shockwave lithotripsy (ESWL) introduced in the 1980s and became the dominant treatment method. This method relies on nonthermal ultrasound therapy mechanisms [27].

Shock wave devices similar to lithotripters are approved for orthopedic indications. The shockwaves are used for treating other problems, such as gall bladder stones, but none have achieved widespread usage. Most lithotripters now are of the electromagnetic design, however very few lithotripters utilize piezoceramic sources. All produce about the same waveform: a 1-μs shocked spike of about 50 MPa followed by a ~10-MPa, 4-μs negative pressure tail. The center frequency might be estimated to be about 150 kHz although. For better treatment efficiency in ESWL treatment, the source is coupled to the patient by a water pillow and transmission gel, and in the remaining original lithotripters through a water bath [12]. About 3000 shock waves are triggered at about 2 Hz repetition rate to pulverize the stone so that the pieces (<2 mm) can pass naturally in urine. Cavitation chips away from the outside, adding cracks that grow by dynamic fatigue and further grind down the stone to passable size [28]. Shockwave Lithotripsy (SWL) works better with some stones than others. Very large stones cannot be treated this way. Stones that are smaller than 2 cm in diameter are the best size for SWL. The treatment might not be effective in very large ones.

Lithotripsy has several important biological side effects, such as blood vessel walls break; Inflammation ensues (i.e., lithotripsy nephritis), scar formation and permanent loss of functional renal mass. SWL is more appropriate for some people than others. Because X-rays and shock waves are needed in SWL, pregnant women with stones are not treated this way. People with bleeding disorders, infections, severe skeletal abnormalities, or who are morbidly obese also not usually good candidates for SWL. In addition to and likely a result of this direct injury cascade, lithotripsy can lead to an accelerated rise in systemic blood pressure, a decrease in renal function, onset of hypertension, an increase in the rate of stone recurrence, and an exacerbation of stone disease. The risks of these adverse bioeffects in lithotripsy have stimulated investigation into mitigation methods with some success. For example, a slower repetition rate (1 Hz) is safer and more effective than the common fast rate (2 Hz) [29], and a pause early in treatment nearly eliminates injury in animals [30]. Overall, lithotripsy has been a therapeutic ultrasound method with a high level of efficacy and patient benefits, but also some important risks particularly for patients requiring repeated treatments. The development of safer treatment protocols for lithotripsy is a prime example of the potential value of research on risk mitigation for optimizing the patient risk/benefit profile in therapeutic ultrasound.

8.2.2.2 Intracorporeal lithotripsy

Intracorporeal lithotripsy to treat urolithiasis means the fragmentation and removal of urinary calculi. These procedures are performed through endoscopes in the urinary tract. Endoscopic lithotripsy refers to the visualization of a calculus in the urinary tract and the simultaneous application of energy to fragment the stone or stones into either extractable or passable pieces. Many calculi in the upper urinary tract are treated with ESWL. However, for stones that are poor candidates for this modality, endoscopic therapy is indicated. Intracorporeal lithotripsy is the favored treatment for many patients, for example, for very large stones, and many different methods and techniques have been reported. The stone may be imaged for guidance by external ultrasound or fluoroscopy, or by ureteroscopic, endoscopic or laparoscopic methods. The number of stones, the size and the composition of the stones and their location are certainly the most important factors in deciding the appropriate treatment for a patient with kidney stones.

Different probes may be used in this method. Electrohydraulic probes which generate a vaporous cavity at the tip have been used in the past (similar to the spark gap external lithotripter but without focusing. The rigid probes may be manipulated percutaneously, but some flexible probes can be applied via the ureter. Rigid ultrasonic probes can utilize both pneumatic action at a few Hz to 1000 Hz, and ultrasonic action at about 25 kHz [31, 32]. Some of bioeffects of intracorporeal lithotripsy are as follows: carries risk of hemorrhage, ureteral perforation, urinary tract trauma, and infection due to the invasive nature of the procedures.

8.2.2.3 Other ultrasound devices (kilohertz-frequency)

A) Harmonic scalpel

The Harmonic scalpel is a surgical ultrasonic instrument used to simultaneously cut and cauterize tissue. It has a 40–80 kHz vibrating titanium rod with a static clamp member, between which the tissue (and blood vessels) is rapidly coagulated due to localized frictional heating [12]. The high-frequency vibration of tissue molecules generates stress and friction in tissue, which generates heat and causes protein denaturation. Also, Harmonic scalpel is widely used in cosmetic surgery for the purpose of removing excessive fat tissue [33]. The mechanism of this invasive method (ultrasound-assisted liposuction), involves cavitational fat cell break up with removal of the fat emulsion by suction through the probe. Unlike electrosurgery, the harmonic uses ultrasonic vibrations instead of electric current to cut and cauterize tissue

B) Sonicators

Sonicators are ultrasonic systems operating in the kHz-frequency regime (20–90 kHz). They are used routinely in biological research to break up cells and tissues, as well as in general and advanced surgical procedures for tissue cutting and hemostasis for tissue removal.

C) Microbubbles and cavitation

Another new technic of ultrasound low frequency therapeutic is microbubble-based therapeutic strategy. In this method, the external ultrasound exposure activates microbubbles (also as drug carriers) at a selected point of treatment in the circulation. Also, microbubble contrast agents are improved the therapeutic efficacy of biologically activated molecules [34]. In general, the enhancement of the concentration of therapeutic biomolecules in the vascular compartment and increased therapeutic agent delivery by extravasation through blood vessels are some mechanisms of this method. Molecules of the therapeutic agent can be incorporated within the bubble shell or loaded into the interior of microbubbles and released in the vascular compartment through ultrasound-induced microbubble disruption [35]. The main advantage of the ultrasound-microbubble based delivery of therapeutic agents over other drug carriers such as nanoparticles or liposomes, is the external control of the ultrasound that used to target the microbubble. The dose of agent to normal tissue is decreased due to a consequent minimization of unwanted drug effects away from the treatment site. The cavitation mechanism is also being exploited to create a new tissue-ablation method known as histotripsy [36]. In histotripsy very high amplitude ultrasound pulses typically of less than 50 μs duration at 750 kHz create a cavitation microbubble cloud to homogenize targeted tissue such as tumors with little heating [37]. Longer HIFU pulses ($t > 3$ ms at 2 MHz) of very high intensity can disrupt tissue by induce rapid heating and generate cavitation and boiling with vapor bubbles [38]. Determining the energy deposited by ultrasound with cavitation as well as the problems of dosimetry and control are challenging. Three important parameters are; understand the medium (including cavitation nuclei), understand the sound field and know when a cavitation effect happens. Passive detection methods, measuring broadband acoustic noise from bubble collapses for monitoring cavitation activity can be deployed and research has indicated useful dosimetric parameters respect to bioeffects [39]. As new cavitation-based treatments are developed, new instrumentation will be needed to reach optimum patient safety. Direct sonothrombolysis using external, typically low-frequency ultrasound has been tested for treatment of thrombotic disease, such as stroke [40]. This new strategy shows promise, but also has shown a potential for deleterious side effects. For example, increased brain hemorrhage was found in a clinical trial for treatment with 300 kHz ultrasound plus tissue plasminogen activator relative to treatment with tissue plasminogen activator alone. Recent work suggests that microbubbles enhance thrombolysis and may be of value in improving stroke therapy [41]. As an example, the transcranial pulsed ultrasound (0.25–0.5 MHz), at relatively low levels (I = 26–163 mW/cm^2), is used to produce cortical and hippocampal stimulation in mice [42]. Since measured temperature gradients were <0.01°C, nonthermal mechanisms for the neuronal effects were hypothesized.

8.2.3 Effects of ultrasonic waves on tissues and blood

The ultrasound waves have different effects on the body. Some can only be justified by linear wave equations. But to explain the cause of some others, one should refer to nonlinear wave relations. Generally, thermal effect, mechanical stress, and bubble stresses are predictable with linear equations. However, in order to provide a mathematical model of secondary effects, such as acoustic force and acoustic power, expansion of the equations to second-order perturbation (nonlinear effect) are required. Passing sound rays on tissue and blood can lead to temperature changes, vibration, and bubble formation. These three major physical factors will continue to be addressed.

8.2.3.1 Temperature rise in the tissue and biofluids

The released ultrasonic waves in a medium are passed, reflected, or absorbed. When the sound waves are absorbed the temperature rises because of the conversion of mechanical energy to heat. For best performance, body temperature should stay at 37°C. Increasing or lowering 1 degree of body temperature only results in a decrease in the efficiency of some parts of body and does not affect the life of cells. By increasing this amount, the time and place of heating should be carefully investigated. The effects of temperature on cell life are shown in Fig. 8.4

At first, the skin temperature is raised due to conversion of mechanical energy to heat in ultrasonic transducers. However, in some cases, it is possible to compensate for the skin's warmth, but it may result in damage, depending on the duration and intensity of the ultrasonic effect. The compensation of the increase in temperature due to the absorption of ultrasonic waves of tissues and internal organs of the body is not possible with the external factor. The blood circulation or heat transfer through tissues has a key role in regulating the body temperature. This mechanism requires a

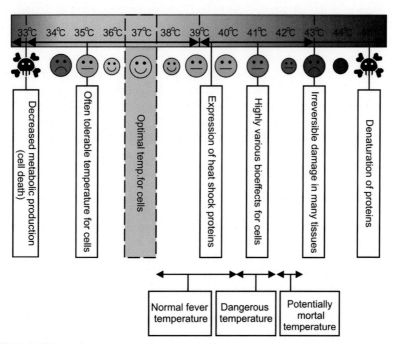

FIGURE 8.4 Effects of temperature change on cells.

special time that is different for each member. If the temperature rise time due to the presence of ultrasound is less than the time it takes for the body to transmit heat, the damage will intensify.

In high intensity focal ultrasonic waves, transmitters with a convergent emission level are used to concentrate the waves to raise the temperature in a particular region. The dimension of this region is proportional to the wavelength (λ). The geometric shape of the focal region is an elliptic and dimensions are obtained from relation (8.1), where K is the geometric characteristic of the audio transmitter [17].

$$1\lambda \times 1\lambda \times K\lambda \tag{8.1}$$

The change in the blood and tissue temperatures during ultrasonic emission are dependent on the characteristics of the transmitter, such as the intensity, duration, and working cycle, as well as tissue characteristics such as the absorption coefficient and perfusion rate, which is modeled by Pennes equation [43]. The thermal index (EMT_{43}) is calculated from the equation relation (8.2) [44]. In Hyperthermia therapy, the radiation is continued until the cancerous thermal index is reached.

$$EMT_{43} = \int_{0}^{t} R^{(T-43)} dt \tag{8.2}$$

8.2.3.2   Bubble formation

One of the causes of bubble formation is the susceptibility to steam or water in the presence of the ultrasonic field [45]. However, bubbles form at frequencies of 20–200 kHz but bubble formation has also been reported at frequencies of about 1 MHz. The bubbles are divided into two categories: stable and unstable. Movement of the stable bubbles creates stresses to adjacent cells, but do not have any harmful effects. In contrast, unstable bubbles that explode and disappear suddenly after several changes in size can damage the adjacent cells [46]. The proper frequency and amplitude of the waves are selected to prevent this adverse event. In other words, in using ultrasonic waves, bubbles do not damage the cells [47]. The mechanical index (MI) for ensuring tissue and blood health and the absence of bubble formation is obtained from Eq. (8.3), which should be less than 0.7 for prevention of bubble formation [44].

$$MI = \frac{p}{f^{0.5}} \tag{8.3}$$

The high-intensity ultrasound waves, the range of 100–10,000 W per square centimeter, the therapeutic and subtherapeutic effects are produced depending on the exposure duration [43]. For example, at intensity levels of the order of 1000 W/cm^2 and exposure durations of the order of 1 s, HIFU produces thermal coagulation within the small volumes determined by the size of the focal spot.

8.2.3.3   Mechanical tension

The ultrasonic beam can cause tension or displacement [48]. The limit of tolerance of tension in any cell or tissue before irreparable damage is clear. If the stress in terms of the amount or the time of impact, more than the amount of tolerance, causes serious damage (e.g., in the red blood cells). In some cases, the forces caused by ultrasound radiation can also move a cell or nerve line. An example of neuronal disorientation has been observed in the mouse brain [47]. One way to reduce the negative effects of tension or displacement is to shift the transmitter over short periods of time. Van Balo [49] reported the effect of shear stress caused by ultrasonic field on cells. He also stated that ultrasound waves may be several times more tolerable than cells. This excessive tension may result in the destruction of the cell. Shear stress due to ultrasonic waves may have destructive biological effects.

8.2.3.4   Nonlinear effects of ultrasound on blood

The ultrasound field has secondary effects on the tissue and blood. Nonlinear terms only reinforce the pressure field in tissue. But the vortex flow creates by them in blood and remains until the end of the radiation time. Also, the acoustic force is applied to the particles. The acoustic radiation force and acoustic streaming are discussed. Due to the high influence of ultrasound on microparticles, acoustic modes can be considered as one of the most effective noncontact methods in biological application [50, 51]. Today, researchers in the field of acoustics call acoustic force as one of the methods for dealing with micronutrients [51]. Settnes and Bruus [51] investigated the forces on a small particle in an ultrasonic field in a viscous flow. In this study, the forces applied to a spherical, compressible, submerged particle in a viscous fluid are analytically calculated by Prandtl-Schlichting boundary theory. It is also claimed that these results are true for any particle diameter and boundary layer thickness smaller than wavelength. Destgeer et al. [52] proposed a method for the stable separation of particles in a microchannel under the influence of moving surface ultrasonic waves. In this method, particles with diameters of 3 and 10 µm, continuously and without contact, are separated by 100% accuracy. Turning off the ultrasonic transmitter will remove all particles from the left path. As the transmitter power increases, the gap between 3 and 10 µm particles increases, and eventually at 151 mW all larger particles are ejected from another path. Muller and et al. [53] investigated the 3D motion of spherical microparticles in a rectangular cross-section channel under the influence of ultrasound waves. They simplified the equations for an isothermal medium and then compared their analytical solution with experimental measurements, which observed a 20% error for a particle with a diameter of half a micrometer.

The steady stream that is created by the absorption of sound in the viscous fluid in the presence of ultrasonic waves is called acoustic streaming [54]. Depending on the mechanism generating the acoustic streaming, different speeds, characteristic lengths and profiles are obtained. Flow velocity range is from about 1 µm/s to 1 cm/s. The flow length scale range is varied from 1 µm to 1 cm [55]. The absorption of acoustic energy in the boundary layer results in acoustic streaming which is much higher than the energy absorption in the other section of the fluid due to the type of acoustic excitation and the aspect ratio Fig. 8.5. Unlike the acoustic flow induced by the boundary layer, Bulk-driven acoustic (Eckart) streaming is the

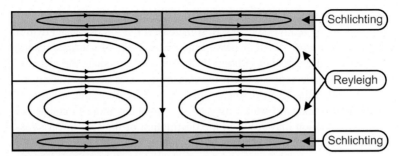

FIGURE 8.5 Gray section, inner boundary layer (Schlichting) and white section, outer boundary layer (Rayleigh).

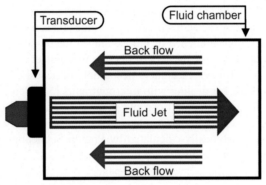

FIGURE 8.6 Eckart flow and backflow.

steady flow resulting from the time-averaged acoustic energy flux density in the bulk of a viscous fluid, Fig. 8.6. The types of currents and their properties are shown in Table 8.4.

Blood in some conditions behaves like a non-Newtonian fluid, which must also be taken into account in modeling. Blood viscosity depends on hematocrit, temperature and strain rate. With constant temperature and blood properties t, the effective parameter is the strain rate which depends on the size of the vessel and the blood flow velocity. Johnston et al. [56] examined non-Newtonian blood flow for the coronary artery with a diameter between 5 and 7 mm. By comparing the stresses at 2, 20, and 100 cm/s velocities, it is found that with increasing velocity, the effect of different non-Newtonian models and their differences with the Newtonian model decreases.

Muller and Bruus [57] examined aspects of the theory of sound and fluid interaction in the microchannel to investigate the effects of viscosity and density on acoustic streaming and ultrasonic force. They calculated the force of sound waves on a spherical particle in a viscous fluid, taking into account the effects of temperature change on the viscosity and density of the fluid. The results show that considering the thermal effects on the viscosity of the fluid causes the acoustic streaming power

Table 8.4 Types of acoustic streaming and specifications [9].

Streaming	Assumptions	Characteristics
Slow	$U_0 \ll u_{19}, Re \ll 1$	$U_0 \propto u_1$ F (acoustic force) $\propto \alpha u_1^2$ $(u \cdot \nabla)u$ (Convective acceleration) ≈ 0
Fast	$U_0 \geq u_{19}, R_e \geq 1$	For coarse-grained streaming, $U_0 \propto \alpha u_1^{2+n}$, n is value for nonlinearity of the wave, $(u \cdot \nabla)u$ (Convective acceleration) Is significant.
Coarse-grained (multidimensional) Eckart	Traveling acoustic wave from single source along an axis $\mathcal{L}$ (characteristic length scale) $\gg \lambda$	Streaming away from sound source into viscous fluid. The Reynolds number determines whether the streaming is fast or slow. $Re_{\mathcal{L}} = \rho_f U_0 \dfrac{\mathcal{L}}{\mu}$
One-dimensional Eckart	$(u \cdot \nabla)u = 0$ Momentum equation is linear. Traveling acoustic wave from single planner source with $\mathcal{L} \gg \lambda$	Streaming away from sound source into unbounded fluid. The Reynolds number determines whether the streaming is fast or slow. $Re_{\mathcal{L}} = \rho_f U_0 \dfrac{\mathcal{L}}{\mu}$
Medium-grained (Rayleigh)	Acoustic wave bounded along sound propagation path: standing acoustic wave. Incompressible everywhere and inviscid in bulk of fluid	The Reynolds number determines whether the streaming is fast or slow. $Re_{\mathcal{L}} = \rho_f U_0 \dfrac{\mathcal{L}}{\mu}$ Vortex flow; the scale of the vortices is approximately the same as the λ. Analysis valid only outside viscous boundary layer.
Fine-grained (Schlichting)	Acoustic wave bounded along sound propagation path: standing acoustic wave. Classical assumptions: incompressible everywhere and inviscid in bulk of fluid.	The Reynolds number determines whether the streaming is fast or slow. $Re_{\mathcal{L}} = \rho_f U_0 \dfrac{\mathcal{L}}{\mu}$ Bounded along sound propagation path: standing wave. Vortex flow; the scale of the vortices is approximately the same as the λ.

U_0, streaming velocity; u_1, acoustic particle velocity.

of the sound flow for 80°C water to be increased by 50% between two rigid parallel plates. For particles of the same density as the fluid, the effects of viscosity on the ultrasound force are negligible, but as the density of particles increases, the viscosity effects increase by up to 30%.

8.3 Governing equations

Despite the consistent principles of acoustic wave equations in physics, the governing equations involving acoustic waves in industry and bioengineering application vary for each type of problem. In fact, the equations needed to analyze the problem are extracted for each case corresponding to its fundamental and conditions. So, first, the acoustic wave equations in general and then the ultrasonic equations, which are acoustic waves that are so high in frequency that humans can't hear them, are introduced.

8.3.1 Linear ultrasonic relationships

Linear perturbations of the ultrasonic field, assuming Newtonian fluid at constant temperature and stationary fluid, are expressed as follow [58].

In the following equations, p, v, ρ, t, c, and u are the acoustic pressure (the local deviation from the ambient pressure), velocity, density, time, the speed of sound, and the vector field particle velocity respectively. Also, terms with only a zero index or a combination of multiplying variables with a zero index constitute the solution of the current before applying the ultrasonic field. Index one indicates the linear perturbations of the waves and terms two indexes represent the secondary and nonlinear effects of the ultrasonic field.

$$p = p_0 + p_1$$
$$\rho = \rho_0 + \rho_1 \tag{8.4}$$
$$\vec{v} = \vec{v}_0 + \vec{v}_1$$

By expanding these perturbations, the continuum equation and the linearized momentum equation for the inviscid fluid are bellow equations:

$$\frac{\partial \rho_1}{\partial t} = -\rho_0 \nabla \cdot \vec{v}_1 \tag{8.5}$$

$$\rho_0 \frac{\partial \vec{v}_1}{\partial t} = -c^2 \nabla \rho_1 \tag{8.6}$$

Eq. (8.7) presents the equation of state.

$$p = p_0 + c_1^2 \rho \tag{8.7}$$

After several times of algebraic simplification, the linear equation of the ultrasonic waves in the ideal fluid is obtained as Eq. (8.8), that Feynman provides for sound in three-dimensions.

$$\nabla^2 P_1 = \frac{1}{c^2} \frac{\partial^2 P_1}{\partial t^2} \tag{8.8}$$

The vector field particle velocity is similar to wave equation:

$$\nabla^2 u = \frac{1}{c^2}\frac{\partial^2 u}{\partial t^2} \tag{8.9}$$

The sound wave equation in one dimension is as follow:

$$\frac{\partial^2 P}{\partial x^2} - \frac{1}{c^2}\frac{\partial^2 P}{\partial t^2} = 0 \tag{8.10}$$

where c and P are the speed of sound and the acoustic pressure (the local deviation from the ambient pressure), respectively.

8.3.2 Westervelt equation

The Westervelt equation is a very popular and high accuracy model for ultrasound modeling. The generalized Westervelt equation is as follow [59, 60]

$$\rho\nabla\cdot\left(\frac{1}{\rho}\nabla p\right) - \frac{1}{c^2}\frac{\partial^2 P}{\partial t^2} + \frac{\delta}{c^4}\frac{\partial^3 p}{\partial t^3} + \frac{\beta}{\rho c^4}\frac{\partial^2 p^2}{\partial t^2} = 0 \tag{8.11}$$

where p, c, δ, β, and ρ are the sound pressure, the speed of sound, the sound diffusivity, the coefficient of nonlinearity and the ambient density, respectively. In Eq. (1), the first term takes diffraction into account. The third term accounts for attenuation. The last term introduces the quadratic nonlinearity. Several simplified versions can be derived from Eq. (1). By using constants for all acoustic parameters, the Westervelt equation for homogeneous media can be recovered [61].

$$\nabla^2 p - \frac{1}{c_0^2}\frac{\partial^2 P}{\partial t^2} + \frac{\delta_0}{c_0^4}\frac{\partial^3 p}{\partial t^3} + \frac{\beta_0}{\rho_0 c_0^4}\frac{\partial^2 p^2}{\partial t^2} = 0 \tag{8.12}$$

where c_0, ρ_0, δ_0, and β_0 are the acoustic parameters for the background medium. The linear acoustic wave equation can be derived by setting β_0 to 0, so that

$$\nabla^2 p - \frac{1}{c_0^2}\frac{\partial^2 P}{\partial t^2} + \frac{\delta_0}{c_0^4}\frac{\partial^3 p}{\partial t^3} = 0 \tag{8.13}$$

Eqs (2) and (3) are useful for studies on characterizing acoustic fields of transducers and for approximately estimating the acoustic field in biological tissue [62]. Eq. (8.11) is used when heterogeneous medium such as skull [59]. On the other hand, nonlinear Eqs (8.11) and (8.12) are used for lithotripsy, histotripsy, and tissue harmonic imaging [63], when high pressure of ultrasound is present. Researchers reported that using the linear acoustic approximation could underestimate the temperature elevation in tissue [62].

8.3.3 Khokhlov-Zabolotskaya-Kuznetsov (KZK) equation

The KZK equation includes diffraction, absorption, and nonlinear effects have been the most widely used model. The KZK equation accounting for energy losses was first published in 1971 [64]. The KZK equation is less accurate in the near field and at a position off the main axis due to be a parabolic approximation of the Westervelt equation. For a focused transducer, the KZK equation is in theory valid for waves traveling within 15–16 degrees of the nominal axis of the beam (typically is the z-axis) [63]. The parabolic form of the wave equation can be solved with efficient numerical techniques, which is the main advantage of the KZK equation. The KZK equation is an equation of evolution type and has the first-order derivative with respect to the propagation main axis.

The KZK equation for thermoviscous and homogeneous material is as follows [63]:

$$\frac{\partial^2 p}{\partial z\, \partial t'} - \frac{c_0}{2}\, \nabla_\perp^2 p - \frac{\delta_0}{2c_0^3}\frac{\partial^3 p}{\partial t'^3} - \frac{\beta_0}{2\rho_0 c_0^3}\frac{\partial^2 p^2}{\partial t'^2} = 0 \tag{8.14}$$

where t' is the retarded time ($t' = t - z/c_0$), $\nabla_\perp^2 = \dfrac{\partial^2}{\partial x^2} + \dfrac{\partial^2}{\partial y^2}$ is the transverse Laplacian.

8.3.4 Kuznetsov equation

The Kuznetsov equation is more accurate than the Westervelt equation. For nondirectional beams, it describes as two forms as follow [64]:

Base on velocity potential (ϕ):

$$\frac{\partial^2 \phi}{\partial t^2} - c_0^2 \nabla^2 \phi = \frac{\partial}{\partial t}\left[\frac{1}{\rho_0}\left(\mu_B + \frac{4}{3}\mu\right)\nabla^2 \phi + (\nabla\phi)^2 + \frac{\beta-1}{c_0^2}\left(\frac{\partial\phi}{\partial t}\right)^2\right] \tag{8.15}$$

where ϕ, μ, and μ_B are the velocity potential, the coefficient of shear viscosity, the coefficient of bulk viscosity, respectively.

Base on acoustic pressure:

$$\nabla^2 p - \frac{1}{c_0^2}\frac{\partial^2 P}{\partial t^2} + \frac{\delta_0}{c_0^4}\frac{\partial^3 p}{\partial t^3} + \frac{\beta_0}{\rho_0 c_0^4}\frac{\partial^2 p^2}{\partial t^2} + \left(\nabla^2 + \frac{1}{c_0^2}\frac{\partial^2}{\partial t^2}\right)L = 0 \tag{8.15}$$

where L is the Lagrangian density of acoustical energy that for plane progressive waves is zero [63].

8.4 Ultrasonic-activated drug delivery

For the past two decades, ultrasonic-activated drug delivery has a major role in cancer therapy. Developed technology helps scientist to improve the delivery of drugs and genes to target tissues while reducing systemic dose and toxicity. In the above section, the physics of US is discussed. In this section, a brief introduction of drug carriers for ultrasound drug delivery and the progress and challenges of ultrasonic-activated drug delivery with special focus on cancer therapy are summarized.

8.4.1   Drug carriers for ultrasound drug delivery

8.4.1.1   Liposomes

Liposome, the most biocompatible nanocarriers that are used for targeted drug delivery, overcome the obstacles of cellular uptake and cause increasing the bio-availability and biodistribution of the compound at the targeted site. Liposomes are self-assembled bilayer phospholipids produced at the temperature above the transition temperature of the lipid by hydration of dry phospholipids in the aqueous medium. Liposome enables to carry both hydrophilic and hydrophobic compounds. Because of this unique property, liposomes are widely used for delivery of chemotherapeutic agents, DNA, proteins, enzymes, etc. [65]. For the development of ligand-targeted liposomes, liposomes are coupled with some functionally active compounds to deliver the therapeutic agent in addition to detect the treatment grade [66]. The drug is loaded in a conjugated form or encapsulated form in liposome. Due to the fact that liposomes reduce the drug toxicity and increase the drug specificity, are known as excellent carriers.

8.4.1.2   Micelles

Micelles consist of polymeric core-shell structures synthesized from amphiphilic block copolymers. Advantages of micelles over other nanocarriers like liposomes are that obtaining liposomes less than 50 nm is difficult and complex [67]. Due to being single-layer structures, the core of the micelles stays hydrophobic, enabling entrapment of poorly soluble drugs in the core of micelles. Also, this feature increases the solubility, improving the bioavailability and circulation time of drugs. High drug loading capacity causes micelles are used for targeted drug delivery [65]. Also, opsonization of micelles with sugar or peptide can result in receptor-mediated targeted drug or gene delivery. Micelles are classified into three categories; self-assembled micelles, unimolecular micelles, and cross-linked micelles.

8.4.1.3   Microbubbles

Microbubbles are gas bubbles ranging between 1 and 8 µm wrapped with protein, peptides, polymer, drugs, etc. Ultrasound imaging and ultrasound-based drug delivery use microbubbles as a contrast agent [34]. Cavitation of microbubbles generates temporary or permanent pores in the blood vessel membrane that increases extravascular delivery of therapeutic compounds to the target organ. Sonoporation allows the delivery of therapeutic compounds without compromising cells' physiological barriers and the defense mechanisms of the cells [65]. As the pores created during sonoporation are temporary, and they usually return to their original confirmation within a few seconds, it is known to be a safe method.

8.4.1.4   Microspheres

Microspheres are polymeric particles ranging from 1 to 1000 µm [65]. Drug can be encapsulated or entrapped form in microsphere in order to drug delivery. Microspheres are used for targeted as well as long-term drug release in the diseased area. Microspheres have four different types: Bioadhesive microspheres that exhibit mucoadhesive property and permits the drug-coated on the surface of the polymer to stick to the targeted organ, resulting in prolonged delivery of the therapeutic agents to the diseased site. Magnetic microspheres which can be used for both diagnostic purposes and drug delivery. The drugs within these particles can be targeted to the diseased area using an external magnetic source [68]. Also, they are commonly utilized for magnetic hyperthermia in tumor tissues. Floating microspheres that are meant to release the drugs loaded in them in gastric content. Radioactive particles which are used for the therapeutic purpose. They directly inject in the veins and link to the targeted organ or tissue.

8.4.2   Sonodynamic therapy (SDT)

Sonodynamic therapy (SDT) is an emerging approach that involves a combination of low-intensity ultrasound and specialized chemical agents known as sonosensitizers. This method has been developed as a novel promising noninvasive approach derived from photodynamic therapy (PDT), that is mediated via ultrasound-induced cavitation and sonosensitizers to produce free radicals to kill dividing cancer cells. The SDT method is used to treat deeply located tumors, however, PDT utilizes visible light, which has limited penetration, and can only be employed superficially or intraoperatively. Jin et al. found that SDT inhibited tumor growth by 77%, compared with 27% for PDT in a subcutaneously located murine squamous cell carcinoma [69]. To improve the efficacy of treating solid tumors, it is important that the sonosensitizer is injected intravenously prior to insonation, rather than directly into the tumor, so that it is more fully and evenly distributed throughout the neoplasm [70].In SDT, the sonication parameters (usually 1.0–2.0 MHz at an intensity of 0.5–3.0 W cm) have been selected to produce inertial cavitation in a cell culture or tumor, where a bubble in a liquid rapidly collapses, producing a shock wave which produces free radicals and a cascade of molecular events that activate the sonosensitizer and in turn damage the cancer cells [71]. Gao et al. reported that SDT also had an antivascular effect and inhibited tumor neovascularization [72]. Combining PDT with SDT had a synergistic effect in solid tumors with additional posttherapy tumor necrosis, inhibition of tumor growth and increased survival times [71]. The combining of the sonosensitizer with a microbubble contrast agent provides new developments in SDT [73].

The sonication parameters in SDT (usually 1.0–2.0 MHz at an intensity of 0.5–3.0 W/cm^2) have been selected to produce cavitation in a cell culture or tumor. Cavitation process involves the nucleation, growth, and implosive collapse of gas-filled bubbles under the appropriate ultrasound conditions. It may be essentially classified into stable and inertial cavitation 14. Bubbles of stable cavitation oscillate, creating a streaming of the surrounding liquid which results in a mixture of the surrounding media while the gas bubbles in inertial cavitation process grow to a near resonance size and expand to a maximum before collapse violently 18. Further, insonation of the microbubble may lead to an additional significant local thermal bioeffects with destruction of the endothelial cells lining the tumor vasculature and a decrease in tumor vascularity [74].

Andrew K.W. Wood and Chandra M. Sehgal grouped the research studies according to the type of cancer cell and the accompanying sonsensitizer that were insonated; to provide a guide to previous sonodynamic studies in which the type of cancer receiving therapy is emphasized [71].

To date the in vivo observations have been performed in implanted subcutaneous tumors of laboratory animals so that there remains a need for future studies in a larger mammal, perhaps using SDT for treating naturally occurring cancers.

8.4.3 Chemotherapy combination with ultrasound

In cancer therapy, there is interest in utilizing low-intensity ultrasound to enhance the delivery of chemotherapeutic agents to a solid tumor. The agents do not selectively target neoplastic cells and thus high levels of the cytotoxic drug in normal tissues [71]. Insonation of a tumor in the presence of the chemotherapeutic agent provides the potential for enhancing the delivery of the agent to the cancer cells whilst minimizing the cytotoxic effects in the contiguous normal tissues.

Different approaches are used to study the ultrasound-mediated chemotherapy. Drug delivery based on chemotherapeutic agents in combination with ultrasound are improved with various techniques that introduced as follows.

8.4.3.1 Chemotherapy in presence of ultrasound (US + Chemo)

Several in vitro and in vivo studies are reported about the potential for low-intensity ultrasound to increase the sensitivity of cancer cells to a chemotherapeutic agent. The human myelomonocytic cells, human uterine carcinoma cells, a human tongue squamous cell carcinoma, a murine lymphoma, murine ovarian cancers are investigated [71]. The frequency and intensity are varied about 0.24–1 MHz and 0.01–7.84 W/cm^2. Both continuous and pulsed waves are used.

8.4.3.2 Chemotherapy in presence of ultrasound and microbubbles (US + Chemo + microbubbles)

Chemotherapy in the presence of microbubbles and ultrasound has the potential to enhance the in vivo delivery of the agent to a tumor with minimum side effects in normal tissues [75]. Watanabe et al. observed that the chemoeffect of cisplatin was enhanced in the presence of ultrasound associated with collapsing and cavitating microbubbles [76]. It should also be noted that the release of the chemotherapeutic agent from the intravascular microbubbles will also impact on the blood vessels, killing the vessels and permitting the therapeutic agent to leave the vessel. The murine colon carcinoma and murine, mammary carcinoma cells, human breast cancer cells, and human myelogenous leukemia cell cancers are some of the several in vitro and in vivo *studies for* chemotherapy in presence of ultrasound and microbubbles. The frequency, intensity, and pressure are varied about 0.5–2.25 MHz, 0.5–4 W/cm^2, and 0.1–2 MPa. Both continuous and pulsed waves are used.

8.4.3.3 Chemotherapy-loaded microbubbles and ultrasound (US + Chemo loaded microbubbles)

The chemotherapeutic drug may be loaded or bound to a microbubble (diameter less than an erythrocyte) [71]. On insonation of a tumor, ultrasound targeted microbubble destruction occurs which has the advantages of being externally controlled and localized to the tumor site. Tinkov et al. reviewed the formation of microbubbles and their potential as carriers for drugs [34]. The structure of the shell of microbubble is important; lipophilic chemotherapeutic drugs including doxorubicin, paclitaxel, and docetaxel can be incorporated into the lipid layer of a microbubble [77]. The concept of a theranostic microbubble which can combine a chemotherapeutic role with ultrasound imaging is receiving increased attention [78]. The bubble is loaded with a chemotherapeutic agent that is released in a tumor under the action of low-intensity ultrasound; the bubbles may also act as a contrast agent and used for contrast-enhanced power Doppler ultrasound imaging of the tumor vasculature. The breast cancer cell and liver cancer in rabbits are two examples of research in this method [72]. The frequency, intensity, and pressure are varied about 0.3–5 MHz, 1–3 W/cm^2, and 0.45–1.2 MPa. Only pulsed waves are used.

8.4.3.4 Chemotherapy-loaded polymeric micelles or liposomes and ultrasound (US + Chemo loaded micelles)

The chemotherapy-loaded polymeric micelles or liposomes is another approach to cancer therapy, that the agent release from the micelle under the action of ultrasound. This method has been used to improve the delivery of chemotherapeutic drugs to a tumor. Staples et al. suggested that cavitation events may have released the doxorubicin from circulating and extravasated micelles into the tumor and insonation could have transiently increased the permeability of the tumor neovasculature [79]. The "shockwave ruptured nanopayload carriers" was developed by Ibsen et al. to minimize the potential side effects a chemotherapeutic agent on normal tissues [80]. The microbubble was encapsulated with a protective outer liposome. The subcutaneous tumor, mice with a subcutaneous breast tumor and a subcutaneous colorectal epithelial cancer cell line in rats are examples of research in this method [71]. The frequency, intensity, and pressure are varied about 0.029–2.25 MHz, 1–5.9 W/cm^2, and 0.173–1.9 MPa. Both continuous and pulsed waves are used.

8.4.3.5 Chemotherapy-loaded liposomes in presence of ultrasound and microbubbles (US + Chemo loaded micelles + microbubbles)

The intracellular absorption of a chemotherapeutic agent is enhanced by inertial cavitation that induced by the insonation of microbubbles, in the other word absorption of ultrasound energy was enhanced by the oscillation and cavitation of microbubbles within the insonated tumor. An intratumoral injection of microbubbles and the intravenous injection of doxorubicin-loaded liposomes was applied by Zhao et al [81]. They found that that cavitation was an important bioeffect of insonation and enhanced the absorption of doxorubicin and the growth of breast cancer tumor in mice was reduced. The mice breast cancer and a murine colorectal adenocarcinoma are investigated by this method [71]. The frequency, intensity, and pressure are set to about 1 MHz, 0.3 W/cm^2, and 1.2 MPa. Only pulsed waves are used.

8.4.3.6 *Chemotherapy-loaded liposomes attached to microbubbles and ultrasound (US + Chemo loaded micelles attached to microbubbles)*

Another method is attaching the drug-loaded liposomes to a microbubble with low-intensity ultrasound. Diagnostic ultrasound is used to confirm the presence of the loaded microbubbles within the tumor vasculature and accurately restricting the delivery of the chemotherapeutic agent to the neoplastic cells by microbubbles. The breast cancer tumors in mice, glioblastoma cells and melanoma cells are studied by this method [71]. The frequency, intensity, and pressure are set to about 1–3 MHz, 2 W/cm^2, and 0.2–0.6 MPa. Only pulsed waves are used.

8.4.3.7 *Chemotherapy in presence of ultrasound and magnetic nanoparticles loaded onto microbubbles and (US + Chemo loaded + magnetic nanoparticle loaded microbubbles)*

Owen et al. and Zhao et al. developed the multimodality imaging, that is, magnetic resonance (MR) imaging in addition to ultrasound imaging; by encapsulation of iron oxide nanoparticles into the microbubble [82, 83]. Niu et al. developed such a multifunctional theranostic agent in which pelvic limb lymph node metastases in rabbits were imaged by both MR and ultrasound (0.3 MHz and 2 W/cm^2 with pulsed waves [84].

   Summary of different approaches for US mediated chemotherapy are shown in Fig. 8.7.

8.4.4   Ultrasound-mediated gene transfection

Ultrasound-mediated gene transfection (sonotransfection) has been developed as a new method for gene therapy, especially for cancer gene therapy. Despite the several advantages of sonotransfection, a low transfection rate is one of the major disadvantages of this method. To improve the transfection rate and the efficiency of sonotransfection, using ultrasound and microbubbles to release or deliver the genes directly into target cells and facilitate gene delivery to neoplastic cells is considered

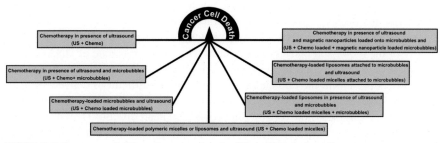

FIGURE 8.7 Summary of different approaches for US mediated chemotherapy.

[71]. Several studies in in vivo and in vitro reported that gene therapy has been successful transfection of genetic material resulting in apoptosis of cancer cells and decreased tumor growth. In some cases, the genetic material and microbubbles were injected separately into a tumor [85] or alternatively they were mixed and then added to the tissue culture or injected into the tumor [71]. The transfection/transduction of genetic material was usually accomplished using nonviral techniques in this method, but using a virus as a carrier with microbubbles was successful too [85, 86]. The nonviral techniques are more safer than viral vector, but has low transfection efficiencies, however, microbubbles are good carriers of genes with a greater capacity for antisense oligonucleotides, and fragment of DNA and even the entire chromosome [71]. Sporadic capillary rupture and increased vascular permeability, as well as enhanced permeability of cell membranes may play a role in the observed efficacy of therapy. Ultrasound-mediated gene transfection is provided a distinct advantage over systemic cancer therapies for tumors with their associated effects on normal tissues. The Melanoma, prostate, squamous cell carcinoma and breast cancers are investigated [71]. The frequency, intensity, and acoustic pressures are varied about 0.021–2 MHz, 0.22–4.6 W/cm^2, and 0.12–2.1 MPa. Both continuous and pulsed waves are used.

8.4.5 Transdermal drug delivery

Transdermal drug delivery (TDD) is the most painless and noninvasive delivery route. In this approach, drug penetrates deep inside the epidermis and dermis applying topically to healthy skin [65]. TDD is the preferable route for vaccination, it improves the bioavailability of the drug and in this method the toxic side effect are minimized [87]. Drug delivery via skin using ultrasound is known as sonophoresis or sometimes called as phonophoresis. Ultrasound increases the porosity of the skin causing enhance TDD. The duty cycle of ultrasound, the distance of horn-to-skin, treatment time, and the composition of the coupling medium are important factor in the efficacy of sonophoresis. Sonophoresis increases the permeability of the skin by thermal effects or by cavitation.

8.4.6 Cardiovascular disease

The microbubble-mediated ultrasound therapy has been studied as a possible tool for cardiovascular diseases. Microbubble destruction provides an increase in local concentration of drug. It releases drug at that particular location and increases the permeability of the biological barriers. Hence it is very useful in a gene or DNA delivery as it eases the transfection process and are used as a delivery vehicle. Because of the barriers present in the endothelium region, it is so difficult to deliver therapeutics in the treatment of atherosclerosis and other cardiovascular diseases like rheumatic heart disease, cardiomyopathy, and congenital heart disease. Treatment in such conditions contains stem cell repair of valve damage and ischemic myometrium, ultrasound-mediated drug delivery through gene therapy, and nanoparticle assisted therapies. Destruction of vascular thrombosis, which is the cause of ischemic stroke using ultrasound waves is called sonothrombolysis. However, ultrasound alone is also capable for sonothrombolysis, but the cavitation due to microbubble enhances the efficacy. High-energy and low-energy ultrasound are used to create sonothrombolysis targeted to monitor the microbubbles due to cavitation into the region of thrombosis, respectively [88]. Also, microbubbles are more capable of holding gases like oxygen and can be used to deliver oxygen. it carries more oxygen as compared to other vehicles and liquid.

8.5   Application of HIFU on thermal ablation

HIFU is widely used in thermal cancer therapy. During sonication, temperature increases in tissues and secondary flow streams in vessels. Tissue temperature rise and blood acoustic streaming mostly depend on ultrasonic field characteristics such as intensity, frequency and pulse duration. In this example, temperature rise due to high intensity focused ultrasonic beams which yields to necrosis of cancerous cells is numerically studied. Solving nonlinear acoustofluidics, second-order of perturbation theory is applied to continuity, momentum, energy, and state equations.

Fig. 8.8 shows the model of tissue, blood vessel, and external ultrasonic source. As seen, 3D geometry is considered. Fully developed laminar blood flow in a 2 mm width and 50 mm length channel assumed. Mean blood flow velocity is considered 1 mm/s before sonication. Cylindrical shape of ultrasonic source concentrates propagating waves. Focal intensity and source frequency in cancerous cells of pancreas tissue are 280 W/cm^2 and 1 MHz respectively. Acoustic absorption coefficients of tissue and blood are frequency dependent.

To solve proposed model wave equations considering viscous terms are necessary. Which are as follow for first and second-order of perturbations:

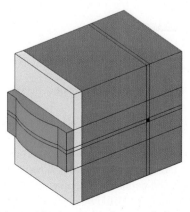

FIGURE. 8.8 The model of tissue, blood vessel and external ultrasonic source.

$$\frac{\partial \rho_1}{\partial t} = -\nabla \cdot \left(\rho_0 \vec{v}_1 \right) \tag{8.17}$$

$$\rho_0 \frac{\partial \vec{v}_1}{\partial t} = -\nabla p_1 + \nabla \cdot \left(\mu_0 \left(\nabla \vec{v}_1 + \nabla \vec{v}_1^{\,T} \right) \right) + \nabla \left(\mu_b \left(\nabla \cdot \vec{v}_1 \right) \right) \tag{8.18}$$

$$p_1 = c^2 \rho_1 \tag{8.19}$$

$$\nabla \cdot \left(\rho_0 \vec{v}_2 + \langle \rho_1 \vec{v}_1 \rangle \right) = 0 \tag{8.20}$$

$$\left\langle \rho_1 \frac{\partial \vec{v}_1}{\partial t} \right\rangle + \left\langle \rho_0 \left(\vec{v}_1 \cdot \nabla \right) \vec{v}_1 \right\rangle = -\nabla \langle p_2 \rangle + \nabla \cdot \left(\mu_0 \left(\nabla \langle \vec{v}_2 \rangle + \nabla \langle \vec{v}_2 \rangle^T \right) \right) + \nabla \left(\lambda_0 \left(\nabla \cdot \langle \vec{v}_2 \rangle \right) \right) \tag{8.21}$$

$$p_2 = c^2 \rho_2 + \frac{1}{2} \frac{\partial c^2}{\partial \rho} \rho_1^2 \tag{8.22}$$

The relations between wave and bio-heat equations in tissue are as below:

$$\rho_t c_{pt} \frac{\partial T_t}{\partial t} - k_t \nabla^2 T_t = Q_{aco} + Q_{bio} \tag{8.23}$$

While blood perfusion effect Q_{bio} is simply calculated by using Eq. (8.23), thermal source of HIFU Q_{aco} should be calculated by solving nonlinear Westervelt equation.

$$Q_{bio} = \rho_0 c_p \omega_b \left(T_2 - T_0 \right) \tag{8.24}$$

$$Q_{aco} = \frac{2\alpha_{abs}}{\omega^2 \cdot \rho_t \cdot cs_t} \left\langle \left(\frac{\partial p_t}{\partial t} \right)^2 \right\rangle \tag{8.25}$$

$$\nabla^2 p_t - \frac{1}{cs_t^2} \frac{\partial^2 p_t}{\partial t^2} + \frac{\delta}{cs_t^4} \frac{\partial^3 p_t}{\partial t^3} + \frac{\beta}{\rho_t cs_t^4} \frac{\partial^2 p_t}{\partial t^2} = 0 \tag{8.26}$$

To solve these equations physical properties of blood and tissue are set as Table 8.5.

Satisfying continuity equation at vessel wall, velocity for second-order perturbations should not be zero.

Boundary conditions for zero and second-order perturbations as well as for first-order perturbations are listed in Tables 8.6 and 8.7, respectively.

As known, noninvasive and painless ultrasonic energy transmission into a tissues is widely used for thermal therapy. HIFU beams do not affect tissues in propagation path. So it is used as a practical method of hyperthermia, which is a major source

Table 8.5 Physical properties values.

Parameters	Definition	Value	SI unit
ρ_{blood}	Blood density	1050[a]	$\dfrac{kg}{m^3}$
ρ_{fat}	Fat density	911[b]	$\dfrac{kg}{m^3}$
ρ_{pan}	Pancreas density	1050[b]	$\dfrac{kg}{m^3}$
cp_{fat}	Heat capacity of fat	2348[a]	$\dfrac{J}{kgK}$
cp_{pan}	Heat capacity of pancreas	3200[b]	$\dfrac{J}{kgK}$
cp_{blood}	Heat capacity of blood	3610[a]	$\dfrac{J}{kgK}$
k_{fat}	Thermal conductivity of fat	0.21[a]	$\dfrac{W}{mK}$
k_{pan}	Thermal conductivity of pancreas	0.51[b]	$\dfrac{W}{mK}$
k_{blood}	Thermal conductivity of blood	0.52[b]	$\dfrac{W}{mK}$
μ_{blood}	Blood viscosity	0.012[c]	Pa s
μ_c	Casson parameters	0.00414[c]	Pa s
τ_c	Casson parameters	0.0038[c]	Pa
α_{blood}	Blood absorption	1.6[b]	$\dfrac{1}{m}$
α_{fat}	Fat absorption	6.4[d]	$\dfrac{1}{m}$
α_{pan}	Pancreas absorption	11.9[b]	$\dfrac{1}{m}$
$c_{s,blood}$	blood sound speed	1578[d]	m s
$c_{s,fat}$	fat sound speed	1440[d]	m s
$c_{s,pan}$	pancreas sound speed	1591[d]	m s
ω_{fat}	Blood Perfusion rate of fat	1.1[a]	$\dfrac{1}{s}$
ω_{pan}	Blood Perfusion rate of pancreas	10[b]	$\dfrac{1}{s}$

[a][89].
[b][90].
[c][91].
[d][92].

Table 8.6 Boundary conditions for zero and second-order perturbations.

	Zero-order	Second-order
Vessel walls	$\overrightarrow{v_0} = 0$	$\overrightarrow{v_2} = \dfrac{-1}{\rho_0}\rho_1\overrightarrow{v_1}$
Vessel inlet	Fully developed velocity $\overrightarrow{v_{0,ave}} = 1\,\dfrac{mm}{s}$	$p_2 = 0$ $T_2 = 0$
Vessel outlet	$p_0 = 0$	$p_2 = 0$ $p_2 = 0$

Table 8.7 Boundary conditions for first-order perturbations.

Transducer surface	$\overrightarrow{a_1} = a\omega^2$
	$T_1 = 0$
Tissue boundary	Perfectly matched layer $T_1 = 0$

of acoustic streaming, too. Concentration of ultrasonic waves in tissue is presented at Fig. 8.9. As seen high intensities occur only at focal area and do not affect other areas.

First-order perturbations of velocity, pressure, and temperature are illustrated at Fig. 8.10.

As v1 is 100 times less than sound velocity, perturbation theory is valid at this step and it can be continued to second order.

Fig. 8.11 is shown the second-order horizontal (u2) and vertical (v2) velocities.

Comparing Figs 8.10 and 8.11, the second-order velocities are 1000 times less than first-order ones. So, assumptions of this theory are totally valid.

Acoustic streaming effect on maximum vessel wall temperature and heat flux is shown in Fig. 8.12A and B, respectively.

Acoustic streaming disturbs boundary layer and local heat flux increases; also rotational flow of blood reduces wall temperature.

As shown in Fig. 8.13, contours show isothermal domain without considering acoustic streaming in microvessel. Blue and red colors present temperature difference while acoustic streaming effects added.

As seen, inside black contours that is focal area no changes occur. But blood temperature close to vessel wall changes up to 50%. High intensity focused ultrasonic waves are able to transfer heat to a specific location at tissue and it is used as a noninvasive method of cancer therapy.

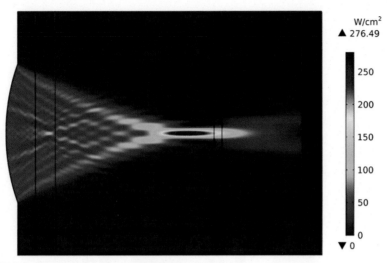

FIGURE 8.9 Concentration of ultrasonic waves in tissue.

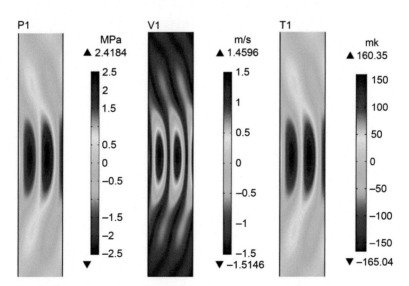

FIGURE 8.10 First-order perturbations of velocity, pressure, and temperature.

Fig. 8.14 is depicted the temperature and ablated volume of tissue. The yellow volume has been reached to temperature above 43°C. This temperature can yield to cell necrosis after 240 min. Three surfaces cut that volume and at red areas sufficient thermal dose to cancer therapy is transmitted.

As shown only at focal area tissue cells have been ablated and others are healthy.

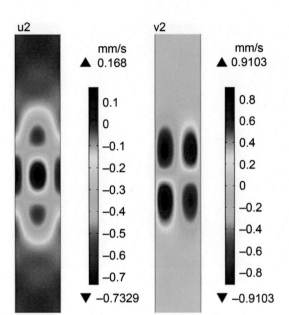

FIGURE 8.11 Second-order horizontal (u2) and vertical (v2) velocities.

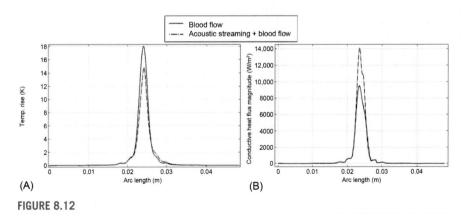

FIGURE 8.12

(A) Temperature and (B) heat flux of vessel wall.

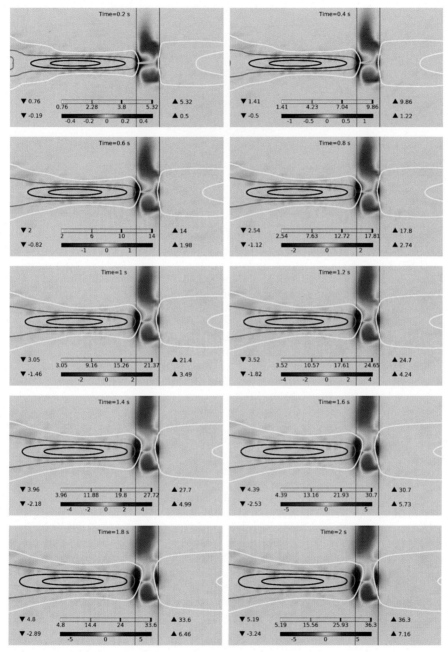

FIGURE 8.13 Isothermal domain and acoustic streaming effects.

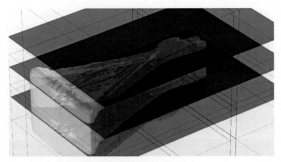

FIGURE 8.14 **Temperature and ablated volume of tissue.**

References

[1] R. Hoogland, Ultrasound Therapy, Enraf Nonius, Delft, (2005).

[2] E. Nussbaum, The influence of ultrasound on healing tissues, J. Hand Ther. 11 (2) (1998) 140–147.

[3] F. Dunn, L.A. Frizzell, Bioeffects of ultrasound, in: J.F. Lehmann (Ed.), Therapeutic Heat and Cold, Rehabilitation Medicine Library, 1982

[4] A. Lenshof, C. Magnusson, T. Laurell, Acoustofluidics 8: applications of acoustophoresis in continuous flow microsystems, Lab Chip 12 (7) (2012) 1210–1223.

[5] C.C. Huang, Y.H. Lin, T.Y. Liu, P.Y. Lee, S.H. Wang, Study of the blood coagulation by ultrasound, J. Med. Biol. Eng. 31 (2) (2011) 79–86.

[6] R. Libgot-Callé, F. Ossant, Y. Gruel, P. Lermusiaux, F. Patat, High frequency ultrasound device to investigate the acoustic properties of whole blood during coagulation, Ultrasound Med. Biol. 34 (2) (2008) 252–264.

[7] M. Wiklund, H. Brismar, B. Önfelt, Acoustofluidics 18: microscopy for acoustofluidic micro-devices, Lab Chip 12 (18) (2012) 3221–3234.

[8] W.G. Pitt, G.A. Husseini, B.J. Staples, Ultrasonic drug delivery–a general review, Expert Opin. Drug Deliv. 1 (1) (2004) 37–56.

[9] J. Friend, L.Y. Yeo, Microscale acoustofluidics: microfluidics driven via acoustics and ultrasonics, Rev. Modern Phys. 83 (2) (2011) 647.

[10] C.C. Chen, P.S. Sheeran, S.Y. Wu, O.O. Olumolade, P.A. Dayton, E.E. Konofagou, Targeted drug delivery with focused ultrasound-induced blood-brain barrier opening using acoustically-activated nanodroplets, J. Control. Release 172 (3) (2013) 795–804.

[11] C.C. Coussios, R.A. Roy, Applications of acoustics and cavitation to noninvasive therapy and drug delivery, Annu. Rev. Fluid Mech. 40 (2008) 395–420.

[12] D.L. Miller, N.B. Smith, M.R. Bailey, G.J. Czarnota, K. Hynynen, I.R.S. Makin, Bioeffects Committee of the American Institute of Ultrasound in MedicineOverview of therapeutic ultrasound applications and safety considerations, J. Ultrasound Med. 31 (4) (2012) 623–634.

[13] J.F. Lehmann, The biophysical basis of biologic ultrasonic reactions with special reference to ultrasonic therapy, Arch. Phys. Med. Rehabil. 34 (3) (1953) 139.

[14] K.G. Baker, V.J. Robertson, F.A. Duck, A review of therapeutic ultrasound: biophysical effects, Phys. Ther. 81 (7) (2001) 1351–1358.

[15] L. Machet, A. Boucaud, Phonophoresis: efficiency, mechanisms and skin tolerance, Int. J. Pharmaceut. 243 (1–2) (2002) 1–15.

[16] L.D. Alexander, D.R. Gilman, D.R. Brown, J.L. Brown, P.E. Houghton, Exposure to low amounts of ultrasound energy does not improve soft tissue shoulder pathology: a systematic review, Phys. Ther. 90 (1) (2010) 14–25.

[17] S.A. Sapareto, W.C. Dewey, Thermal dose determination in cancer therapy, Int. J. Radiat. Oncol. Biol. Phys. 10 (6) (1984) 787–800.

[18] C. Marchal, Clinical trials of ultrasound hyperthermia, Ultrasonics 30 (1992) 139–141.

[19] J.E. Kennedy, G.R. Ter Haar, D. Cranston, High intensity focused ultrasound: surgery of the future? Br J. Radiol. 76 (909) (2003) 590–599.

[20] Y. Zhou, High-intensity focused ultrasound treatment for advanced pancreatic cancer, Gastroenterol. Res. Pract. (2014) 2014.

[21] A. Gelet, J.Y. Chapelon, R. Bouvier, O. Rouviere, Y. Lasne, D. Lyonnet, J.M. Dubernard, Transrectal high-intensity focused ultrasound: minimally invasive therapy of localized prostate cancer, J. Endourol. 14 (6) (2000) 519–528.

[22] S. Thüroff, C. Chaussy, G. Vallancien, W. Wieland, H.J. Kiel, A. le Duc, et al. High-intensity focused ultrasound and localized prostate cancer: efficacy results from the European multicentric study, J. Endourol. 17 (8) (2003) 673–677.

[23] B. Larrat, M. Pernot, J.F. Aubry, R. Sinkus, M. Tanter, M. Fink, Radiation force localization of HIFU therapeutic beams coupled with magnetic resonance-elastography treatment monitoring in vivo application to the rat brain, in: 2008 IEEE Ultrasonics Symposium, IEEE, 2008, pp. 451–454.

[24] I.R.S. Makin, T.D. Mast, W. Faidi, M.M. Runk, P.G. Barthe, M.H. Slayton, Miniaturized ultrasound arrays for interstitial ablation and imaging, Ultrasound Med. Biol. 31 (11) (2005) 1539–1550.

[25] R. Chopra, K. Tang, M. Burtnyk, A. Boyes, L. Sugar, S. Appu, L. Klotz, M. Bronskill, Analysis of the spatial and temporal accuracy of heating in the prostate gland using transurethral ultrasound therapy and active MR temperature feedback, Phys. Med. Biol. 54 (9) (2009) 2615.

[26] S.E. Jung, S.H. Cho, J.H. Jang, J.Y. Han, High-intensity focused ultrasound ablation in hepatic and pancreatic cancer: complications, Abdominal Imag. 36 (2) (2011) 185–195.

[27] A.Z. Weizer, P. Zhong, G.M. Preminger, New concepts in shock wave lithotripsy, Urol. Clin. N. Am. 34 (3) (2007) 375–382.

[28] O.A. Sapozhnikov, A.D. Maxwell, B. MacConaghy, M.R. Bailey, A mechanistic analysis of stone fracture in lithotripsy, J. Acoust. Soc. Am. 121 (2) (2007) 1190–1202.

[29] K.T. Pace, D. Ghiculete, M. Harju, R.J.D.A. HONEY, University of Toronto Lithotripsy AssociatesShock wave lithotripsy at 60 or 120 shocks per minute: a randomized, double-blind trial, J. Urol. 174 (2) (2005) 595–599.

[30] R.K. Handa, M.R. Bailey, M. Paun, S. Gao, B.A. Connors, L.R. Willis, A.P. Evan, Pretreatment with low-energy shock waves induces renal vasoconstriction during standard shock wave lithotripsy (SWL): a treatment protocol known to reduce SWL-induced renal injury, BJU Int. 103 (9) (2009) 1270–1274.

[31] S.C. Kim, B.R. Matlaga, W.W. Tinmouth, R.L. Kuo, A.P. Evan, J.A. McAteer, et al. In vitro assessment of a novel dual probe ultrasonic intracorporeal lithotriptor, J. Urol. 177 (4) (2007) 1363–1365.

[32] G. Lowe, B.E. Knudsen, Ultrasonic, pneumatic and combination intracorporeal lithotripsy for percutaneous nephrolithotomy, J. Endourol. 23 (10) (2009) 1663–1668.

[33] Mann, M. W., Palm, M. D., & Sengelmann, R. D. (2008, March). New advances in liposuction technology. In Seminars in cutaneous medicine and surgery (Vol. 27, No. 1, pp. 72-82). WB Saunders.

[34] S. Tinkov, R. Bekeredjian, G. Winter, C. Coester, Microbubbles as ultrasound triggered drug carriers, J. Pharmaceut. Sci. 98 (6) (2009) 1935–1961.

[35] K. Ferrara, R. Pollard, M. Borden, Ultrasound microbubble contrast agents: fundamentals and application to gene and drug delivery, Annu. Rev. Biomed. Eng. 9 (2007) 415–447.

[36] K. Kieran, T.L. Hall, J.E. Parsons, J.S. Wolf, J.B. Fowlkes, C.A. Cain, W.W. Roberts, Refining histotripsy: defining the parameter space for the creation of nonthermal lesions with high intensity, pulsed focused ultrasound of the in vitro kidney, J. Urol. 178 (2) (2007) 672–676.

[37] Z. Xu, M. Raghavan, T.L. Hall, M.A. Mycek, J.B. Fowlkes, C.A. Cain, Evolution of bubble clouds induced by pulsed cavitational ultrasound therapy-histotripsy, IEEE Trans. Ultrasonics Ferroelectr Freq. Control 55 (5) (2008) 1122–1132.

[38] M.S. Canney, V.A. Khokhlova, O.V. Bessonova, M.R. Bailey, L.A. Crum, Shock-induced heating and millisecond boiling in gels and tissue due to high intensity focused ultrasound, Ultrasound Med. Biol. 36 (2) (2010) 250–267.

[39] J.H. Hwang, J. Tu, A.A. Brayman, T.J. Matula, L.A. Crum, Correlation between inertial cavitation dose and endothelial cell damage in vivo, Ultrasound Med. Biol. 32 (10) (2006) 1611–1619.

[40] R.J. Siegel, H. Luo, Ultrasound thrombolysis, Ultrasonics 48 (4) (2008) 312–320.

[41] K.E. Hitchcock, C.K. Holland, Ultrasound-assisted thrombolysis for stroke therapy: better thrombus break-up with bubbles, Stroke 41 (10 (Suppl 1)) (2010) S50–S53.

[42] Y. Tufail, A. Matyushov, N. Baldwin, M.L. Tauchmann, J. Georges, A. Yoshihiro, S.I. Helms Tillery, W.J. Tyler, Transcranial pulsed ultrasound stimulates intact brain circuits, Neuron 66 (5) (2010) 681–694.

[43] E.S. Ebbini, G. Ter Haar, Ultrasound-guided therapeutic focused ultrasound: current status and future directions, Int. J. Hypertherm. 31 (2) (2015) 77–89.

[44] J. Ventura, K.H. Nuechterlein, J.P. Hardesty, M. Gitlin, Life events and schizophrenic relapse after withdrawal of medication, Br. J. Psychiatry 161 (5) (1992) 615–620.

[45] R. Mettin, A.A. Doinikov, Translational instability of a spherical bubble in a standing ultrasound wave, Appl. Acoust. 70 (10) (2009) 1330–1339.

[46] J. Rooze, E.V. Rebrov, J.C. Schouten, J.T. Keurentjes, Dissolved gas and ultrasonic cavitation—a review, Ultrasonics Sonochem. 20 (1) (2013) 1–11.

[47] M. Wiklund, Acoustofluidics 12: Biocompatibility and cell viability in microfluidic acoustic resonators, Lab Chip 12 (11) (2012) 2018–2028.

[48] J. Wang, J. Dual, Theoretical and numerical calculation of the acoustic radiation force acting on a circular rigid cylinder near a flat wall in a standing wave excitation in an ideal fluid, Ultrasonics 52 (2) (2012) 325–332.

[49] E. VanBavel, Effects of shear stress on endothelial cells: possible relevance for ultrasound applications, Prog. Biophys. Mol. Biol. 93 (1–3) (2007) 374–383.

[50] T.G. Jensen, Acoustic Radiation in Microfluidic Systems (Doctoral dissertation, Master's thesis), Technical University of Denmark, Department of Micro and Nano Technology, 2007.

[51] M. Settnes, H. Bruus, Forces acting on a small particle in an acoustical field in a viscous fluid, Phys. Rev. E 85 (1) (2012) 016327.

[52] G. Destgeer, K.H. Lee, J.H. Jung, A. Alazzam, H.J. Sung, Continuous separation of particles in a PDMS microfluidic channel via travelling surface acoustic waves (TSAW), Lab Chip 13 (21) (2013) 4210–4216.

[53] P.B. Muller, M. Rossi, A.G. Marin, R. Barnkob, P. Augustsson, T. Laurell, C.J. Kähler, H. Bruus, Ultrasound-induced acoustophoretic motion of microparticles in three dimensions, Phys. Rev. E 88 (2) (2013) 023006.

[54] J. Lighthill, Acoustic streaming, J. Sound Vib. 61 (3) (1978) 391–418.

[55] P.B. Muller, Acoustofluidics in Microsystems: Investigation of Acoustic Streaming, DTU Nanotech, Department of Micro-and Nanotechnology, (2012) (Master's thesis).

[56] B.M. Johnston, P.R. Johnston, S. Corney, D. Kilpatrick, Non-Newtonian blood flow in human right coronary arteries: steady state simulations, J. Biomech. 37 (5) (2004) 709–720.

[57] P.B. Muller, H. Bruus, Theoretical aspects of microchannel acoustofluidics: thermoviscous corrections to the radiation force and streaming, Proc. IUTAM 10 (2014) 410–415.

[58] H. Bruus, Acoustofluidics 2: perturbation theory and ultrasound resonance modes, Lab Chip 12 (1) (2012) 20–28.

[59] Y. Jing, T. Wang, G.T. Clement, A k-space method for moderately nonlinear wave propagation, IEEE Trans. Ultrasonics Ferroelectr. Freq. Control 59 (8) (2012) 1664–1673.

[60] G.F. Pinton, J. Dahl, S. Rosenzweig, G.E. Trahey, A heterogeneous nonlinear attenuating full-wave model of ultrasound, IEEE Trans. Ultrasonics Ferroelectr. Freq. Control 56 (3) (2009) 474–488.

[61] Y. Jing, M. Tao, G.T. Clement, Evaluation of a wave-vector-frequency-domain method for nonlinear wave propagation, J. Acoust. Soc. Am. 129 (1) (2011) 32–46.

[62] I.M. Hallaj, R.O. Cleveland, FDTD simulation of finite-amplitude pressure and temperature fields for biomedical ultrasound, J. Acoust. Soc. Am. 105 (5) (1999) L7–L12.

[63] J. Gu, Y. Jing, Modeling of wave propagation for medical ultrasound: a review, IEEE Trans. Ultrasonics Ferroelectr. Freq. Control 62 (11) (2015) 1979–1992.

[64] V.P. Kuznetsov, Equations of nonlinear acoustics, Sov. Phys. Acoust. 16 (1971) 467–470.

[65] B. Joshi, A. Joshi, Ultrasound-based drug delivery systems, Bioelectronics and Medical Devices, Woodhead Publishing, 2019, pp. 241–260.

[66] L. Sercombe, T. Veerati, F. Moheimani, S.Y. Wu, A.K. Sood, S. Hua, Advances and challenges of liposome assisted drug delivery, Front. Pharmacol. 6 (2015) 286.

[67] V.K. Mourya, N. Inamdar, R.B. Nawale, S.S. Kulthe, Polymeric micelles: general considerations and their applications, Indian J. Pharm. Educ. Res. 45 (2) (2011) 128–138.

[68] A. Joshi, S. Solanki, R. Chaudhari, D. Bahadur, M. Aslam, R. Srivastava, Multifunctional alginate microspheres for biosensing, drug delivery and magnetic resonance imaging, Acta Biomater. 7 (11) (2011) 3955–3963.

[69] Z.H. Jin, N. Miyoshi, K. Ishiguro, S.I. Umemura, K.I. Kawabata, N. Yumita, et al. Combination effect of photodynamic and sonodynamic therapy on experimental skin squamous cell carcinoma in C3H/HeN mice, J. Dermatol. 27 (5) (2000) 294–306.

[70] K. Ninomiya, C. Ogino, S. Oshima, S. Sonoke, S.I. Kuroda, N. Shimizu, Targeted sonodynamic therapy using protein-modified TiO_2 nanoparticles, Ultrasonics Sonochem. 19 (3) (2012) 607–614.

[71] A.K. Wood, C.M. Sehgal, A review of low-intensity ultrasound for cancer therapy, Ultrasound Med. Biol. 41 (4) (2015) 905–928.

[72] Z. Gao, J. Zheng, B. Yang, Z. Wang, H. Fan, Y. Lv, H. Li, L. Jia, W. Cao, Sonodynamic therapy inhibits angiogenesis and tumor growth in a xenograft mouse model, Cancer Lett. 335 (1) (2013) 93–99.

[73] Y. Zheng, Y. Zhang, M. Ao, P. Zhang, H. Zhang, P. Li, L. Qing, Z. Wang, H. Ran, Hematoporphyrin encapsulated PLGA microbubble for contrast enhanced ultrasound imaging and sonodynamic therapy, J. Microencapsul. 29 (5) (2012) 437–444.

[74] B.J. Levenback, C.M. Sehgal, A.K. Wood, Modeling of thermal effects in antivascular ultrasound therapy, J. Acoust. Soc. Am. 131 (1) (2012) 540–549.

[75] C.H. Heath, A. Sorace, J. Knowles, E. Rosenthal, K. Hoyt, Microbubble therapy enhances anti-tumor properties of cisplatin and cetuximab in vitro and in vivo, Otolaryngol.–Head Neck Surg. 146 (6) (2012) 938–945.

[76] Y. Watanabe, A. Aoi, S. Horie, N. Tomita, S. Mori, H. Morikawa, Y. Matsumura, G. Vassaux, T. Kodama, Low-intensity ultrasound and microbubbles enhance the antitumor effect of cisplatin, Cancer Sci. 99 (12) (2008) 2525–2531.

[77] J. Kang, X. Wu, Z. Wang, H. Ran, C. Xu, J. Wu, Z. Wang, Y. Zhang, Antitumor effect of docetaxel-loaded lipid microbubbles combined with ultrasound-targeted microbubble activation on VX2 rabbit liver tumors, J. Ultrasound Med. 29 (1) (2010) 61–70.

[78] S.A. Peyman, R.H. Abou-Saleh, S.D. Evans, Research spotlight: microbubbles for therapeutic delivery, Therapeut. Deliv. 4 (5) (2013) 539–542.

[79] B.J. Staples, W.G. Pitt, B.L. Roeder, G.A. Husseini, D. Rajeev, G.B. Schaalje, Distribution of doxorubicin in rats undergoing ultrasonic drug delivery, J. Pharmaceut. Sci. 99 (7) (2010) 3122–3131.

[80] S. Ibsen, M. Benchimol, D. Simberg, S. Esener, Ultrasound mediated localized drug delivery, in: Nano-Biotechnology for Biomedical and Diagnostic Research, Springer, Dordrecht, 2012, pp. 145–153.

[81] Y.Z. Zhao, D.D. Dai, C.T. Lu, H.F. Lv, Y. Zhang, X. Li, et al. Using acoustic cavitation to enhance chemotherapy of DOX liposomes: experiment in vitro and in vivo, Drug Dev. Ind. Pharm. 38 (9) (2012) 1090–1098.

[82] J. Owen, Q. Pankhurst, E. Stride, Magnetic targeting and ultrasound mediated drug delivery: benefits, limitations and combination, Int. J. Hypertherm. 28 (4) (2012) 362–373.

[83] Y.Z. Zhao, L.N. Du, C.T. Lu, Y.G. Jin, S.P. Ge, Potential and problems in ultrasound-responsive drug delivery systems, Int. J. Nanomed. 8 (2013) 1621.

[84] C. Niu, Z. Wang, G. Lu, T.M. Krupka, Y. Sun, Y. You, et al. Doxorubicin loaded superparamagnetic PLGA-iron oxide multifunctional microbubbles for dual-mode US/MR imaging and therapy of metastasis in lymph nodes, Biomaterials 34 (9) (2013) 2307–2317.

[85] Q. Tang, X. He, H. Liao, L. He, Y. Wang, D. Zhou, et al. Ultrasound microbubble contrast agent–mediated suicide gene transfection in the treatment of hepatic cancer, Oncol. Lett. 4 (5) (2012) 970–972.

[86] F. Li, L. Jin, H. Wang, F. Wei, M. Bai, Q. Shi, L. Du, The dual effect of ultrasound-targeted microbubble destruction in mediating recombinant adeno-associated virus delivery in renal cell carcinoma: transfection enhancement and tumor inhibition, J. Gene Med. 16 (1–2) (2014) 28–39.

[87] A. Joshi, J. Kaur, R. Kulkarni, R. Chaudhari, In-vitro and ex-vivo evaluation of raloxifene hydrochloride delivery using nano-transfersome based formulations, J. Drug Deliv. Sci. Technol. 45 (2018) 151–158.

[88] E. Unger, T. Porter, J. Lindner, P. Grayburn, Cardiovascular drug delivery with ultrasound and microbubbles, Adv. Drug Deliv. Rev. 72 (2014) 110–126.

[89] H.P. Kok, A.N.T.J. Kotte, J. Crezee, Planning, optimisation and evaluation of hyperthermia treatments, Int. J. Hypertherm. 33 (6) (2017) 593–607.

[90] M.S. Adams, S.J. Scott, V.A. Salgaonkar, G. Sommer, C.J. Diederich, Thermal therapy of pancreatic tumours using endoluminal ultrasound: parametric and patient-specific modelling, Int. J. Hypertherm. 32 (2) (2016) 97–111.

[91] S. Karimi, M. Dabagh, P. Vasava, M. Dadvar, B. Dabir, P. Jalali, Effect of rheological models on the hemodynamics within human aorta: CFD study on CT image-based geometry, J. Non-Newton. Fluid Mech. 207 (2014) 42–52.

[92] H. Azhari, Doppler Imaging Techniques, 2010.

Application of microfluidics in cancer treatment

9

Chapter outline

9.1 Introduction

Microfluidic technology can precisely control and manipulate a small amount of fluid on the microscale, typically submillimeter in a confined and limited environment. It integrates multiple processes into a small chip that normally requires a lot of laboratory equipment [1,2]. Multidisciplinary field such as engineering, physics, chemistry, biochemistry, nanotechnology, and biotechnology joints together to design systems in which low volumes of fluids are processed to achieve multiplexing, automation and high-throughput screening. A microfluidic platform provides a set of fluidic unit operations that is done for an easy combination within a well-defined fabrication technology. Microfluidic platform clears a generic and continuous way for miniaturization, integration, automation, and parallelization of biochemical processes [3]. To reach a small volume of liquids, micropumps generate a flow to circulate liquids on an artificially constructed channel of various geometries (microchip) [4]. Also micromixers

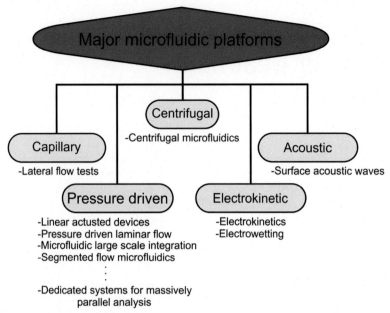

FIGURE 9.1 Classification of microfluidic platforms.

are mixed or moved small volumes of liquids in microchips [4]. These microfluidic devices will be discussed in the following sections. Microfluidic platforms are classified into five groups according to the main liquid propulsion principle: capillary, pressure driven, centrifugal, electrokinetic, and acoustic systems (Fig. 9.1) [1].

Characterization of the respective platforms is provided in Table 9.1 [1]. The surface acoustic wave (SAW) as an important technology in cancer treatment is described in the following.

9.1.1 Surface acoustic waves

SAWs are ultrasonic waves that propagate on an elastic surface. When acoustic waves come in contact with the fluid environment, they can be coupled to the fluid and become a fluid flow stimulus. This effect is known as acoustic streaming. The SAWs have been employed to move droplets on chemically modified or functionalized piezoelectric substrate (Fig. 9.2) [4]. Exciting the interdigital transducers (IDTs) which are deposited onto the piezoelectric substrate causes droplets moving. A certain high-frequency voltage is applied to excite the sound waves [4]. The SAW devices are easy to construct and operate, also they are powerful platforms to direct chemical sensing, upstream microfluidic processing and sample preparation [5].

The application of SAW technology in microfluidics which is an important tool, originally discovered by Lord Rayleigh in 1880. He observed when an earthquake occurs there is a surface wave and the ground begins to roll. Transverses and

Table 9.1 Characterization of the respective microfluidic platforms.

Microfluidic platform	Characterization
Lateral flow tests	The liquids are driven by capillary forces in lateral flow tests (test strips) such as pregnancy test strip. Liquid movement is controlled by the wettability and feature size of the porous or microstructured substrate.
Linear actuated devices	In this method the liquid movement is controlled by mechanical displacement of liquid such as a plunger. Mostly limited to a one-dimensional liquid flow in a linear fashion without branches or alternative liquid pathways are used.
Pressure driven laminar flow	The hydrodynamically stable laminar flow profiles in microchannels is achieved by a pressure driven laminar flow platform, characterized by liquid transport mechanisms based on pressure gradients. The different implementations in terms of using external or internal pressure sources such as using syringes, pumps, or micropumps, gas expansion principles, pneumatic displacement of membranes are developed.
Microfluidic large scale integration	A microfluidic channel circuitry with chip-integrated microvalves based on flexible membranes between a pneumatic control channel layer and a liquid-guiding layer, introduces a microfluidic large-scale integration. By applying pneumatic pressure to the control-channels, the microvalves are closed or open. The more complex units such as micropumps, mixers, and multiplexers are built up on one single chip by combining several microvalves.
Segmented flow microfluidics	The small liquid plugs and/or droplets immersed in a second immiscible continuous fluid as stable micro-confinements within closed microfluidic channels are discussed as a segmented flow microfluidics. Their volume range are in the picolitre to microliter.They can be merged, split, sorted, and processed without any dispersion in microfluidic channels.
Centrifugal microfluidics	The frequency protocol of a rotating microstructured layer are controlled all processes in centrifugal microfluidics. The liquid is transported by important forces such as centrifugal, Euler, Coriolis, and capillary forces.
Electrokinetics	The electric fields or electric field gradients acting on electric charges or electric dipoles, respectively, are controlled by the electrokinetics platforms microfluidic. Several electrokinetic effects such as electroosmosis, electrophoresis, dielectrophoresis, and polarization superimpose each other. Electroosmosis can be used to transport the whole liquid bulk while the other effects can be used to separate different types of molecules or particles within the bulk liquid.
Electrowetting	Electrowetting platforms as stable micro-confinements use submerged droplets in a different immiscible phase (gas or liquid). The droplets reside on a hydrophobic surface that contains a one- or two-dimensional array of individually addressable electrodes. The wetting behavior is defined by voltage between a droplet and the electrode under the droplet. By changing voltages between neighboring electrodes, the droplets behavior is controlled.
Surface acoustic waves	The surface acoustic wave platform uses droplets residing on a hydrophobic surface in a gas. The microfluidic unit operations are mainly controlled by acoustic shock waves traveling on the surface of the solid support. The shock waves are generated by an arrangement of surrounding sonotrodes, defining the droplet manipulation area. Most of the unit operations such as droplet generation are freely programmable.

FIGURE 9.2 IDT placed on the piezoelectric substrate generates SAWs.

longitudinal forms of waves are found in solids. Rayleigh waves cause a particle in a solid to move in an elliptical path along the surface and much energy is not dissipated into the air because of boundary conditions of the earth's surface. The SAWs using a piezoelectric substrate coupled to IDT. The rapid changes in the electric field can generate ultrasonic waves similar to Rayleigh waves [4].

9.1.2 Generation of SAW-induced streaming

A part of traveling SAW that has contacted liquid, reflect in the form of the longitudinal wave into the liquid because liquid viscosity increases relative to the substrate. When acoustic refraction occurs, the changed SAW is called leaky SAW. The IDT excites SAW and the SAW propagates along the piezoelectric substrate surface within the depth of a single wavelength. The refraction angle is known as a Rayleigh angle and refracted wave moves along this direction. The refracted longitudinal waves generate a force in their propagation direction and induce flow into the limited liquid. The boundaries of the limited liquid reflect the actuated liquid and lead to internal streaming. Transformation of the SAW attenuation into a steady fluid flow (as a nonlinear phenomenon) is called SAW-induced acoustic streaming. Small amplitudes of SAW is suitable for nano and micromixing. The internal streaming in the small fluid volume is shown in Fig. 9.3. At large amplitudes of SAW, droplets transportation and streaming of closed fluid layers at the surface of the substrate occurs [6–9].

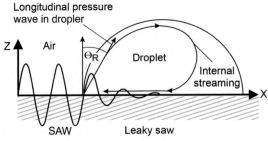

FIGURE 9.3 A small droplet on a piezoelectric substrate exposed to the acoustic streaming.

Applying mechanical vibration as a unique property in SAW introduced a variety analytical technologies. For example, a SAW sensor is used to analyse and detect the changes in the environment. Also SAW-induced microfluidics is used to actuate a liquid on the surface of the piezoelectric substrate, and this actuation is used for various processes [5].

9.1.3 Application of SAW

Regard to the recent researches, SAWs can be an effective means of fluid control and as a particle in lab-on-a-chip devices [10]. SAW technology onto lab-on-a-chip platforms has opened a new frontier in microfluidics. These advantages cause SAW microfluidics to get involved in various biology, chemistry, engineering, and medicine application [11]. An efficient actuation of fluids on the microscopic scale such as mixing, pumping, atomizing and driving, as well as the dexterous manipulation of micro-objects (cells, droplets, particles, nanotubes, etc.) such as separation, sorting, trapping, concentration, merging, patterning, and focusing in open (sessile droplets) and confined spaces (microchannels/chambers) are required to realization of microscale total analysis systems (μTASs) and lab-on-a-chip technologies. The powerful acoustofluidic technics based on high frequency (10–1000 MHz) SAWs are used to achieve this objective. The SAW-based miniaturized microfluidic devices are best known for their noninvasive properties, low costs, and the ability to manipulate micro-objects in a label-free manner. Acoustofluidic technics are classified according to the use of traveling SAWs (TSAWs) or standing SAWs (SSAWs). The schematic of these waves is shown in Fig. 9.4. TSAWs are used to actuate fluids and manipulate micro-objects via the acoustic streaming flow (ASF) as well as the acoustic radiation force (ARF). The SSAWs are mainly used for micro-object manipulation and are rarely employed for microfluidic actuation [12].

TSAWs are used to achieve fluid mixing, fluid translation in open space, microfluidic pumping in enclosed channels, jetting and atomization, particle/cell concentration, droplet, and cell sorting, and reorientation of nano-objects. It shows that TSAWs are a key component for many emerging on-chip applications.

In SSAW-based devices, instead of harnessing the acoustic streaming, the primary ARFs which act on particles via the surrounding fluid are used. The SSAW-based devices apply these primary ARFs to focus a flow stream of particles into a single-file line, separate a flow stream of particles based on particle properties, actuate a single particle/cell moving with a flow stream, pattern a group of particles in stagnant fluid,

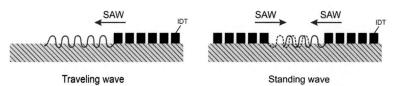

FIGURE 9.4 The schematic of the traveling SAWs (TSAWs) and standing SAWs (SSAWs).

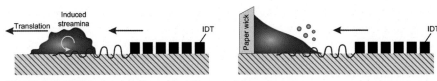

FIGURE 9.5 SAW device couples with a fibrous wick such as paper.

manipulate a single particle/cell/organism in stagnant fluid, manipulate proteins, and align micro/nanomaterials [11].

In addition to fluids and particles manipulation and control by SAW microfluidic devices, they can be used for sensitive detection and sensing. Microsensors based on SAW can be integrated with other SAW-based microfluidic components [13,14]. Because of these properties, it is possible to realize SAW-based, fully integrated, true lab-on-a-chip systems that can be launched into practical settings [11]. The SAW devices that use for sensing purpose, operate at high frequencies (~100–500 MHz), but most microfluidic devices work at lower frequencies (~10–100 MHz) [5] SAW-based sensors, especially for biological targets, often operate with liquid on the surface of the SAW device [5]. A SAW microfluidic device can be utilized in conjunction with other technologies of sensing and detecting to analyze minute volumes of liquid samples. The SAW wave that is reflected in the liquid bulk causes multiple phenomena inducement in the Liquid [5]. Another SAW microfluidics approach is that SAW device couples with a fibrous wick such as paper or thread and makes continuous solution deliver to the SAW surface from some external reservoir (as shown in Fig. 9.5) [5].

As well, SAWs have been used to sprays and aerosols generation for nearly two decades. Several new applications have been developed such as spray cooling131 for thermal management, spray coating, and aerosolizing particles to load an optical trap [15]. However, in chemical analysis and lab-on-a-chip applications, two primary applications have developed [5]. Although SAWs often are not utilized for detection technic, they provide an effective, low power, and relatively simple means to conduct many upstream steps necessary for effective analysis [5].

The applicability of SAW-based actuators for various microfluidic tasks include particle separation, fluid mixing, localized heating, and fluid atomization [16]. Because of inadequate electrical impedance in IDTs, absorption of SAW power in vessel walls or incompatible wavelengths, the energy efficiency in lab setups incorporating SAW-based microfluidic actuators is often insufficient. Furthermore, an inefficient mode of operation can severely damage the device. Regarding the intended microfluidic task, several optimization strategies can be combined that cause significant improvements in the device performance. Essential parameters for the functionality of SAW-based actuators are the efficiency of the energy conversion to the manipulated system and the available electrical power from the signal source. In real-world systems, input power supplied by the signal source is reduced by losses in the acoustic or electric regime and it causes limitation of the SAW power for actuation.

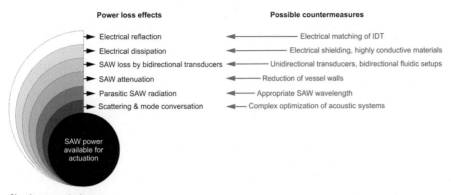

FIGURE 9.6 The most important power loss effects with a brief overview of suitable countermeasures.

Fig. 9.6 shows the most important power loss effects together with a brief overview of suitable countermeasures. Increasing the power delivered by the high-frequency signal source is the simplest and efficient approach to counteract the power loss. However, an increase in the power available by high-frequency sources results in cost increases. Additionally, an inefficient device operation due to the power loss can increase the local temperature of microfluidic devices, which can lead to decreased system performance, especially for portable devices. In portable devices and integrated high-power devices including SAW atomizers, acoustic tweezers, and mixers, the power that source provides should be used with the highest possible efficiency to minimize costs, system size, and temperature-associated device degradation.

It is possible to do different reactions against the power loss that leads to the extension of the applicability and performance of SAW actuators into the field of hand-held devices, portable devices, and devices with high power need. In the following section the microfluidic systems are more explained.

9.2 Microfluidic system

Since the motion of fluids is involved in most of biological assays and biological procedures, the emergence of microfluidics has become a useful tool for biological applications and analysis. Due to the advantages of microfluidic, they are used in a variety of application including detection and separation, small-scale chemical and micro-/nano-particle synthesis, single-cell biology, drug delivery, cell culture, DNA chips, lab-on-a-chip technology, micro-propulsion, and micro-thermal technologies, coculture system for studying the interactions between tumor cells, or bacteria system, or neighboring normal cells for characterizing the effects of new drugs, clinical diagnostic applications, assisted reproductive technology (e.g., in vitro fertilization) [17,18], studying resistance to chemotherapy, screening, and optimizing drug combinations, monitoring cellular responses, study of cellular, tissue and total-body behavior upon irradiation, microbubbles production, etc. [19,20]. Also microbubbles can be used in gene therapy and as a drug delivery vehicle for cancer treatment through antivascular effects [20]. In order to therapeutic targets finding and new drug testing for cancer, diverse experimental models of types in vivo, ex vivo, and in vitro have been traditionally employed [21].

Microfluidic devices provide highly controlled environments and that can be configured in different cell-culture formats such as single-cell culture and automatic cell culture [19]. In vitro cell culture with microfluidic devices, provides a high temporal resolution in single-cell level quantification of cellular responses to different stimuli. Microfluidic technics have been utilized to look at linear DNA specifically using restriction endonucleases. The fragments of DNA within the microfluidic channels can be functionalized with fluorescent proteins. By these technics, the analysis of how control of DNA modifications, the genetic contents of DNA, and the sizes of DNA fragments via proteins using small volumes of analysis is possible [22].

Immunoassays are another of technic which has been optimized by microfluidic. Immunoassays are used to measures the presence or concentration of a macromolecule or a micromolecule in a solution by using a tagged antibody or an antigen which emits a detectable signal. Recently, microfluidic methods have been developed to optimize this time-consuming process, by shortening the reaction times and washing steps [22]. The polymers such as polydimethyl sulfoxide (PDMS), molecules such as self-assembled monolayers (SAM) have been used for the fabrication of microchannels to functionalize substrates with certain chemical compounds to immobilize proteins, DNA or cells [23,24]. For example, the microchannels are made of PDMS which would minimize the diffusion distances by replenishing the diffusion layer with molecules [25]. Multivalent ligand binding studies have been performed by using a PDMS glass/microfluidic chip. Supported lipid vesicles with the incorporated ligand were analyzed within these channels. The lipid compound 2,4 dinitrophenyl (DNP) was fused onto the glass surface to form a continuous lipid bilayer. First, a fluorescently labeled anti-DNP with predetermined concentrations were injected into these microchannels. Then by using total internal reflection fluorescence microscopy (TIRFM) the surface-bound antibodies were imaged [25].

The PDMS microchannels with good biocompatibility have been applied for cell-based assays. For example, PDMS was constructed with two inlets and various staggered channels which were used as chaotic mixers to serially dilute the molecules which are connected to the main channel. A PDMS gradient generator was employed to study the effects of chemotaxis. To monitor cell migration, the concentration gradients of interleukins were administered onto a neutrophil substrate at the main channel. The migratory characteristics of the neutrophil shifted to the region of the highest gradient of interleukins [25]. Also the PDMS is used to control fluid flow for deposit cells and proteins onto the microfabricated channels. The criteria of microfluidic cell separation technology are listed in Table 9.2.

Table 9.2 Criterions of microfluidic cell separation technology.

Method	Sample	Separation way	Separation resolution	Throughput
Mechanical filters	Human metastatic cells spiked in whole blood	Size exclusion	$17 \pm 1.5\ \mu m$ CTCs separated from most smaller blood components	0.75 mL/min, $\sim 10^9$ cells/min
Hydrodynamic	Diluted blood (0.3%), liver cells $2\text{-}3 \times 10^6$/mL Diluted blood (50%)	Streamline manipulation	–	$\sim 10^5$ cells/min
Inertia	Diluted blood ($\sim 2\%$), cells, blood and bacteria	Lift forces and secondary flows	$\sim 5\ \mu m$	$\sim 10^6$ cells/min
Gravity	Microparticles	Sedimentation differences	$\sim 17\ \mu m$	$\sim 17\ \mu L$/min
Acoustic	Whole blood	Acoustic radiation force	$>1\ \mu m$	80 μL/min 10^8 cells/min
Optical	Silica and protein microparticles	Optical lattice	$<0.54\ \mu m$	2.4×10^3 particles/min

9.2.1 Microfluidic devices

The pumps, micromixers, and valves are such microfluidic devices that have been utilized for technics such as immunoassays, PCR, electrophoresis, cell counting, and cell sorting and will be discussed in the following [22]:

9.2.1.1 Valves

Valves are integrated into the silicon microchip to control flow rates. Various passive and active micromixers have been fabricated for such applications. External devices are used to provide the energy of active microvalve and actuation control. For example, the hydrogels are used to active micromixer that changes in pH or temperature causes they can polymerize. The gelling of the hydrogels in microscale devices can be controlled by diffusion of the protons or heat convection of the stream layer to regulate the flow [26]. Passive micromixers use a temporary flow stop device and limit the flow to one direction. Silicon or elastomers is used in the construction of passive one-way valves in order to direct fluid flow.

9.2.1.2 Pumps

Pressure-driven flows are generated by syringe pumps which allow molecules carried in a fluid stream and diffused throughout the microchannel. For pressure regulation, these pumps require special tubing and cumbersome micropump devices. The

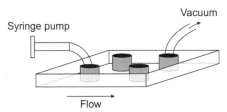

FIGURE 9.7 The schematic of the syringe pump connected to a PDMS microchannel.

electroosmotic pump is a more appropriate device which applies an electric field to liquid movement. If the walls of a microchannel have an electric charge, an electric double layer of counter ions will form at the walls [27]. Fig. 9.7 shows the schematic of the syringe pump connected to a PDMS microchannel. The ions in the double layer move toward the opposite polarity electrode by applying an electric field across the channel. This causes the movement of fluid near the walls and convective motion of the bulk fluid occurs because of viscous forces. The fluid flow can be controlled by adjusting the voltage. Usually high voltages are required in electroosmotic pump. If a specific protein or molecule adds to the walls of the microchannel, the fluid flow can alter due to their charges [27].

9.2.1.3 Micromixers

Micromixers are necessary for a variety of applications particularly in the field of biology and chemistry. The basic concept of mixing involves diffusion, which changes the distribution of particles as well as the concentration gradient in an aqueous solution. The laws of Brownian motion explain that random motion occurs when the forces are imposed on spherical beads in a water solution and pushes them to the left and right creating an imbalance [28].

In macroscopic scales, turbulence causes the random variation in fluid flow and dispersion of the molecules resulting in the mixing of particles along with the fluid flow. In microscopic scales, creating turbulence is a daunting task, thus micromixers specifically designed to cause disturbances or alterations to fluid flow are incorporated into microchips. In microfluidic systems, two categories of micromixing methods are employed to induce a more rapid mixing either by passive or active mixing. In passive mixing, obstacles are created to multiple fluid streams to fold the liquid layers that enhance diffusion in a defined area of the channel. Indeed, passive mixing relies on diffusion. Some designed micromixers are included parallel lamination, where a basic design of a microchannel with a geometry consisting of a T or Y shaped mixer (Fig. 9.8).

Active micromixers utilize different external energy sources to achieve dynamic mixing. These micromixers are categorized based on the types of external sources as pressure field driven, electrical field driven, sound field driven, magnetic field driven, and thermal field driven. In electrohydrodynamic mixing, two platinum wires are implemented and placed perpendicular to the mixing channel that induces mixing by changing the voltage and frequency. In magnetohydrodynamic disturbance, an

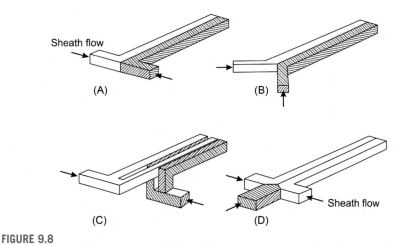

FIGURE 9.8

The different models of the parallel lamination micromixer: (A) the basic T-mixer and (B) Y-mixer, (C) the concept of parallel lamination, and (D) the concept of hydraulic focusing.

external magnetic field is needed to apply DC voltages on the electrodes to induce a mixing movement in the chamber [28,29] and in thermal disturbances, thermal energy enhances mixing of the fluid layers. Therefore, heating devices are used to regulate temperature causing thermal energy.

9.2.1.4 Types of materials

Material properties basically have an impact on functionality and production capability. Successful design and fabrication require an appropriate selection of material. As well, if the material selection is suitable, it causes to balance functional requirements such as biocompatibility, mechanical resilience, optical transparency, and chemical resistance. For fabrication, the behavior of material determines which processes are favorable, unfavorable, or impossible [30].

Generally, materials that are used for microfluidic devices are categorized into polymers, silicones, glass, and metals. The use of ceramics, composites, and other materials is less common [30]. Properties that should be attended in microfluidic devices manufacturing as follow:

1. Mechanical properties: The mechanical properties are more important for microfluidic devices, such as flexures, flaps, membranes, and other structures, in which deformation is an essential part of their functionality. Mechanical properties are commonly dependent on temperature, though the degree of dependency is variable and relates to the type of material. By increasing temperature, most of the materials tend to exhibit higher ductility and lower stiffness and thermoplastics being most extreme in these regards.
2. Thermal properties: For microfluidic devices, thermal properties are important from several aspects. Geometric design parameters such as

microchannel cross section and electrode spacing can significantly be changed if materials are exposed to substantial thermal expansion. The design of some microvalves and micropumps are based on thermomechanical actuation. Microfluidic reactors need the ability to set up and control the temperature of the reactants. Monitoring sensitive changes in heat transfer is necessary in order to effective heat transfer that is the basic purpose for microchannel electronics cooling, and some gas sensors. Fluid main properties such as viscosity are temperature dependent, therefore thermal properties of microfluidic components are also principle issues. The thermal properties are important from a production standpoint, even if there are no temperature-related issues for design functionality. The application of heat to thermoplastic materials for geometric shaping is so important. For multistep fabrication processes, the complete sequence of thermal conditions must also be considered.

3. Electrical properties: For functional purposes such as electrostatic actuation, electrokinetic pumping, and capacitive sensing, the electrical properties of materials are important. Some of the most essential properties are resistivity, relative permittivity (i.e., dielectric constant), and breakdown voltage.

4. Optical properties: Transparency over a wider range of the spectrum beyond visible light is particularly important for devices based on optical methods [31,32]. Laser-induced fluorescence is a highly sensitive detection method.

5. Magnetic properties: Ferromagnetic materials are significantly influenced by magnetic fields that are specially evidenced by force interaction. Nickel, cobalt, and iron are common metals that exhibit ferromagnetism. Nickel and nickel alloys are common microfabrication due to established processes.

6. Other physical properties and characteristics: Other physical properties that should be considered are density, gas permeability, and water absorption. Hydrophobicity is not a material property because of high dependence on surface conditions but it is also wildly relater to microfluidics.

9.2.1.5 Considerations

Several functions should be integrated within a compact platform in lab-on-chip technology that cause to be easily transportable and can deliver rapid data output.

Approximately, surface condition relates to all aspects of microfluidics. Plasma treatment, chemical surface treatments and laser irradiation are used on this subject. Usually surface conditions effect significantly on microfluidic functionality [30].

Temperature is one of the functions is required to be controlled from two aspects, in terms of profile (homogeneous or gradient) and the accessible range, that is possible with the greatest accuracy. The regulation of temperature is one of the important parameters manages physical, chemical, and biological applications. Some applications need tight control of temperature such polymerase chain reaction (PCR), temperature gradient focusing for electrophoresis (TGF), digital microfluidics, mixing, and protein crystallization [33]. Maintaining the temperature at 37°C to keep cells alive and controlled temperature cycling for PCRs are principle requirements in biological applications. Spatial homogeneity of the temperature profile is desirable in both cases [34].

PCR is widely used in molecular biology to amplify target DNA in vitro [34]. As well, the generation of temperature gradients to drive droplets is another approach. Indeed, droplet-based microfluidics is a viable alternative to manage small volumes (typically a few hundreds of a nanoliter). Generally, thermal regulation has found application in a wide range of various fields such as the development of efficient and rapid mixing technics, or the screening of solubility diagrams to study protein crystallization. The external heating and integrated heating are two ways that heat diffuses from sources such as microwave or laser, toward the liquid either.

9.2.1.6 Cell washing

Cell and bead washing are the main experimental method which has extensive usage in biomedical research and biological studies. Cell/bead washing is an essential sample preparation method used in diverse cell studies and analysis. Advantages of standing surface acoustic wave (SSAW)-based washing device include label-free manipulation, simplicity, high biocompatibility, high recovery rate, high washing efficiency, and compatibility with other on-chip components. It is useful for lab-on-a-chip applications [35,36].

As well, bead washing has been regularly used in molecular biology and immunology. For example, multiple bead-washing steps are required for reagents changing when affinity-based DNA purification is performed using QIAEX® beads [37].

Commonly, the centrifugation method is used for cells or beads washing. To overcome the limitations of centrifugation-based cell washing methods such as low biocompatibility and its difficulty for in-line integration, which is necessary to realize automatic, µTAS, microfluidic technics have been developed to wash cells and beads in a continuous flow. In order to have more control of the movement of cell/bead during the washing process, significant efforts have been made to use external forces to manipulate cells/ beads, leading to the development of several active cell washing technics [38,39]. External forces such as magnetic forces, dielectrophoretic (DEP) forces, or acoustic forces are applied to cells/beads flowing in these devices [40]. As a consequence, cells/beads are extracted from their original medium stream and placed into a wash solution. Of all these methods, acoustic methods offer remarkable advantages in terms of label-free manipulation, biocompatibility, and versatility.

Recently, SSAWs have been used to accomplish label-free manipulation of various micro/nano-objects such as beads, cells, droplets, microorganisms, and nanowires [41,42]. Exclusively, a unique formation of tilted-angle standing surface acoustic wave (taSSAW) has been introduced that the IDTs are inclined at a specific angle to the flow direction [43]. In comparison with previous SSAW approaches where the IDTs are aligned in parallel with the flow direction [44,45].The taSSAW approach presents a significantly larger lateral displacement for particles. Consequently, cells/beads can be effectively separated due to differences in size, density, and/or compressibility. The potential application of taSSAW approach has been investigated in the development of a SSAW-based cell/bead washing device. By optimizing the input voltage, beads washing with a high recovery rate (>98%) and high washing efficiency (>97%) have been done. Preparation of WBCs from whole blood through RBC lysis is widely employed in immunology and clinical diagnosis [46,47] and purification of white blood cells (WBCs) from lysed blood samples is investigated.

By connecting polydimethylsiloxane (PDMS) microchannel to a piezoelectric substrate, SSAW-based cell/bead washing device is made. A pair of IDTs is placed on both sides of the microchannel. SAWs are generated from both IDTs when a radio frequency (RF) signal is applied to IDTs. The SAWs diffuse in opposite directions on the substrate surface and leak into microchannel containing liquid. A SSAW field is organized due to the interference between them that causes pressure fluctuations in the liquid. As a result, a periodic distribution of pressure nodes and pressure anti-nodes is formed inside the microchannel. The ARF and Stokes drag force are exerted to particles flowing into the SSAW field.

Due to the competition of the acoustic radiation and Stokes drag forces, the cells or beads flowing into the SSAW field will deviate from their original medium stream. Thus, cells/beads can be washed out from the original medium and collected through the outlet. The different microfluidics based single-cell analysis including cellular analysis, single-cell manipulation, genetic analysis, protein analysis, and single-cell analysis using flow cytometry have been investigated [48]. All of these procedures have some advantages and disadvantages and can be combined according to device application in order to device performance improvement.

The use of microfluidic technology has many advantages as follow:

1. capability to analyze single molecules such as proteins, DNA, and eventually to single cells [4].
2. capacity for high-throughput screening.
3. precise control of flow due to the presence of microfluidic channels which allows the study of many important biomechanical processes such as shear stress [21].
4. high efficiency.
5. potential to achieve higher sensitivities.
6. capability to process extremely low volumes of samples and reagents, which significantly reduce the cost of assays and this is particularly important in the case of irradiated patient samples which are precious as the collection is usually difficult.
7. the fast analytical times.
8. portability for point-of-care applications.
9. enabling excellent spatial control of cell distribution at physiological length scales.
10. ability to control the local cellular microenvironment precisely, without (or with less) interference from the external environment that causes better monitoring of cellular behavior.
11. facilitating high parallelization of experiments.
12. less contamination.

13. improvement of reproducibility.

14. eliminating the need for skilled technicians.

15. enabling smooth and predictable transport of materials through the device via laminar flow because of narrow channels and preventing turbulent flow.

By contrast microfluidic technology usage for modeling purpose have some limitation [21]:

1. if subsequent biochemical studies are needed retrieving cells from microfluidic devices is possible but not straightforward.

2. If a large number of cells are required for downstream biochemical studies or if secreted proteins should be quantified, the low cell number can also be a drawback.

3. because device materials in terms of mechanical properties (e.g., typically PDMS or glass coverslip) are much stiffer than the typical hydrogels in which cells are embedded, might also constitute an issue for some 3D studies. Hence, only cells that are far from the PDMS or glass surfaces should be analyzed.

4. PDMS can adsorb small hydrophobic molecules and it can cause problems for drug screening studies.

5. lack of organ specificity is the current limitation of in vitro metastatic models.

9.3 Microfluidic systems in cancer

The obvious potential of microfluidic technologies to help the advance of cancer research has been characterized. They are especially important in metastatic cascade studying, due to the manner of metastatic spread that occurs through several steps and it is difficult to resolve them in vivo. Containing metastasis by conventional therapies have limited success and it causes 90% of cancer-related deaths. Intravital imaging is so useful in metastasis study [49,50], but it needs specialized expertise and equipment. As well, in several visceral organs because of their deep location, access to high image resolution is hard [21]. In addition, some events of the metastatic cascade such as intravasation are rare and hard to image in vivo, but this rate of occurrence can be more readily modulated in vitro. Each step of the cascade represents a promising target for therapeutic intervention. Progression of solid tumors can be divided into steps, include an epithelial-to-mesenchymal transition (EMT), invasion, intravasation, transport of circulating tumor cells (CTCs) in the bloodstream, extravasation in the distant organ, and recolonization at the metastatic site [51]. Microfluidic assays have been designed to repeat definite features of some of these steps, such as EMT [52], cancer cell invasion and adhesion [53], intravasation and extravasation [54,55].

In vitro metastatic models have some limitations that lack of organ specificity is one of the main current limitations. The importance of this issue is because there are developing evidence that shows cancer cells relate to their microenvironment [21], particularly with organ-specific cells which leads to organ selectivity in metastasis [56]. Therefore, targeted therapeutic approaches for inhibiting metastasis can be achieved by understanding organ-specific interactions. Now, microfluidics is the most appropriate in vitro systems that mimic organ-specific environments since they exactly control the spatial distribution of different cell types mimicking the in vivo settings. To date, studies have added organ-specific cell types or chemokines to microfluidic models and replicated certain aspects of organ architecture [57]. In this context, these models such as the addition of different immune cells for studying their role in metastasis are developed [21]. Newly, reducing metastasis in melanoma or kidney cancer patients is reported by using immunotherapies [58].

Tissue biopsy and rebiopsy used to cancer diagnosis and metastasis monitoring until now is a very invasive procedure that is limited only to certain locations and not always possible in clinical practice [59]. It is difficult or impossible to obtain the tissue sample through tissue biopsy in some cases as well. Information about a very small area of tumor is yielded at extraction time by tissue biopsy [59]. It can pose a risk to the patient because of invasive nature and it has significant cost [60].

There is another problem that biopsies usually suffer from sample bias due to heterogeneity of tumor and is caused false diagnosis and it is the main reason for the failure of cancer treatment because of incapability to capture the heterogeneity during tumor development. Tumors are dynamic and their mutation pattern changes that resists nonspecific treatment and it causes problems to patients undergoing targeted therapies. Liquid biopsy is a new diagnostic concept which can help to eliminate these problems. By this method, the circulating tumor cells (CTC) and cell-free circulating tumor DNA (ctDNA or cfDNA) released into the peripheral blood during metastasis in terms of therapeutic targets and drug resistance-conferring gene mutations is analyzed.

Liquid biopsy can be used as a cancer diagnostic toolkit which provides an opportunity for screening, monitoring, and treatment response and recurrence detection after surgery [61]. Undoubtedly, liquid biopsy is a valuable addition to the oncologist's tool because it improves disease monitoring over time and avoids painful procedures such as conventional tissue biopsy and radiological assessment. As well, in order to tumor burden monitoring in the blood and early detection of emerging resistance of targeted cancer therapies, liquid biopsy can be used. Circulating tumor cells (CTCs), a component of the "liquid biopsy", are generally accepted as an indicator for early cancer diagnosis [62]. Many new technologies have been developed to the detection of CTCs in the peripheral blood of patients with solid epithelial tumors (e.g., breast, prostate, lung, and colon cancer). Detection of CTCs has been extensively reported in various metastatic cancers including breast, prostate, lung, and colorectal cancers. Recent research demonstrates, because of the existence of the significant relation between CTC enumeration and prognosis of cancer patients, CTCs are either surrogate of the metastatic activity or causally involved in the metastatic process [59]. Dielectrophoresis technic is another method, based on hydrodynamic flow, that has been used for target CTCs separation. In this method, dielectric forces apply through microelectrode arrays and cells that get separated based on similar properties [63]. The efficiency of this method was reported by more than 90%. In addition, blood flow should be slow in this method, and being in an isotonic medium with low conductivity. Also, CTC isolation using the dielectrophoresis technic is reported in recent studies [64].

Cancer cell separation using acoustic waves (acoustofluidics) is a new way that has been reported in the past few years. An ultra-high-throughput acoustophorestic microdevice, that was able to remove RBCs from human whole blood with 95% efficiency, was described by Adams et al. As well, acoustofluidic technology can be used to separate prostate cancer cells from WBCs in blood. Yang and Soh have utilized acoustofluidics in MCF7 breast cancer to sort of viable cells from nonviable [65,66]. Nevertheless, cell separation based on acoustic-wave is still in the early stages of research and needs more development and optimization [59]. Novel progress technologies will able the molecular characterization of CTCs and the definition of biomarkers for therapeutic strategies and knowledge of the process of cancer metastasis. Thus implementing CTC analyses as a liquid biopsy using microfluidic lab-on-a-chip medical devices can give new insights into anticancer drug mechanisms.

In recent years, many microfluidic platforms in order to CTC detection and isolation based on size were presented. The performance of a microfluidic device which enables to capture LNCaP-C4-2 prostate cancer cells with efficiency more than 95%, in such a way that operates under constant pressure at the inlet for blood samples. It creates a uniform pressure differential across all the microchannels in the array. Optimization of rows to achieve a capture of more than 95% demonstrated that trapping chambers with five or six rows of microconstriction is needed. The shape of cancer cells deformed in the constriction channel. The blood flow temporarily decreased because of cancer cell deformation [62].

CTC measurement is a cancer screening method with a minimum invasion that is useful for the early staging of cancer patients and in monitoring recurrent or metastatic disease. In order to attain decisive results for CTC detection and enrichment, generally, ~7.5 mL of blood volume is needed due to rare enumeration of CTCs in the blood (1–100 cells/mL). A limitation of the current CTC enumeration systems approved by the US Food and Drug Administration (FDA) namely CellSearchR and AdnaTestR. According to the result of clinical studies, CTC detection in blood in cases of breast cancer, colorectal cancer, and prostate cancer patients can be done by CellSearch [62].

The necessity of having minimal sample preprocessing is ideal for CTC isolation and enrichment that can avoid loss of CTC and it should have high throughput, high efficiency, high sensitivity, high purity, and low cost. Because of considering these criteria and in order to eliminate the limitations associated with surface-marker-based approaches, the microfluidics has been motivated to develop new CTC technologies based on the biophysical properties of CTCs. These rely on the assumption that biophysical attributes such as size, deformability, permittivity, and conductivity of CTCs are different from blood cell properties. Microchips based on size exclusion, deformability, and dielectrophoresis have been explored and shown promising results in CTC collection [67,68]. Also, SAWs that are generated by tilted identical IDTs on microfluidic channels were utilized to CTCs collection of whole blood. For CTC enrichment, finding of balance between capture efficiency, throughput, and

Table 9.3 Some of the important existing technologies about CTC.

Devices	Key observations (recovery rate, throughput, purity)
Micropillar chip; Herringebone chip	5-1281 CTCs/mL with 1 mL/h, detection rate 65%; 63 CTCs/mL with 1.2 mL/h, detection rate 91.8 ± 5% for PC3; 50.3 CTCs/mL with 8 mL/h, detection rate 98% SKBR3;
Micropillar chip	30 CTCs/mL with 1 mL/h, detection rate 97 ± 3%
Microfilter	0–12.5 CTCs/mL with 90 mL/h, detection rate 90%
Microchannel array	10 CTCs/mL with 36 µL/min, recover rate >95%. Using additional grooved surface can achieve over 90% capture rate with >84% purity. Throughput: 3.6 mL/h.
µHall: antibody labeled magnetic nanoparticles	Recovery rate of MDA-MB-468 99%; purity 100%; throughput 3.25 mL/h; analyzing ovarian 7.6 CTCs/mL.
Portable filter	MCF-7, SK-BR-3: 5 CTCs/mL at 0.5 psi, constant pressure, detection rate >90%
Microsieve filter with uniform pore structure	Undiluted whole blood with flow rate < 2 mL/min. MCF-7 recover rate >80%; accurately detected CTC from 8 patients.
Lab-on-disc: centrifugal microfluidic size selection	MCF-7: 61% capture rate; have the potential to high-throughput and easy detection by directly reading on disc. After collection, the MCF-7 still viable in 15 days cell culture. Detect 0–90 CTCs/7.5 mL. Throughput: >3 mL/min; sensitivity: 95.9 ± 3.1% recovery rate; selectivity: >2.5log WBCs depletion.

purity lead to the development of microfluidic chip [62]. Some of the important existing technologies about CTC are listed in Table 9.3.

CTC high-throughput entrapment chip (CTC-HTECH) is one of the newest microfluidic devices based on the size and deformability of CTC that have multiple rows of microconstriction channels and trapping chambers. The limitations of inefficient capture that were observed in existing size-dependent CTC capture designs have been eliminated using CTC-HTECH. The blood cells in blood flow that includes blood cells and CTCs pass through the microchannel, while single CTC is captured in trapping chambers. The number of CTCs is determined when trapping chambers is scanned. This ability is not available in other CTC microfluidic chips based on size. The cells were then detached from the flask with a trypsin-EDTA solution (Sigma Aldrich). The size of prostate cancer cells is in the range of 10–15 µm which is larger than blood cells [62]. The schematic of the flow configuration for the acoustophoresis cell separation is shown in Fig. 9.9.

The use of a syringe pump to the creation of constant flow can guarantee the throughput of microfluidic devices which are designed with a limited number of microchannel. The pressure will be redistributed as cells are trapped in the microchannel by using the constant flow rate to a device with a large quantity of microchannel in an array. So the mechanical drag forces on the trapped cells within different channels are varied [62]. The increased pressure will cause the cell deformation that could damage cells, leading to poor recovery of viable cells [69,70]. The alteration in

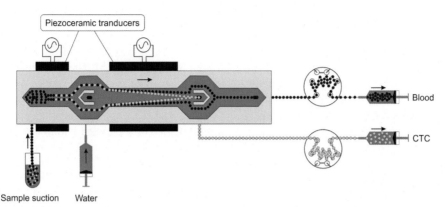

FIGURE 9.9 The schematic of the flow configuration for the acoustophoresis cell separation.

cytoskeleton strength, membrane stiffness, and the biomechanical properties occur by a sudden change in pressure. By contrast, if a pressure pump is used in a constant pressure mode, alteration of the backpressure causes pressure supply and it is maintained constant. A variation of the flow occurs because of trapped cells in the microchannel but excessive pressure is avoided that results to not impose additional mechanical stress on the cells. In a microfluidic system with a low Reynolds number, the constant pressure mode will apply constant mechanical force on the cell membrane. Due to the fact that CTC capture and enrichment depends on biomechanical attributes and CTCs properties, a gentle change in both pressure and flow rate is required. In CTC-HTECH device programmable pressure pump was used in order to provide constant pressure to the blood samples at the inlet of device that caused a low-stress environment for CTC trapping and enrichment [62].

Size is a main biophysical property of cells that has been used to CTC separation from blood cells. Different microfluidic chips in PDMS and the other biocompatible polymers have been designed and fabricated to separate larger epithelial tumor cells from smaller blood cells [71]. In size-based trapping method, attainment to favorable efficiency and purity of CTC trapping depends on the selection of geometry and dimension. In microfilters, for example, the array of small holes with a size around 8 μm allows most blood cells to pass through but captures CTCs [72].

Recently, S. Lee's group has studied a flow-restricted microfluidic channel with an array of trapping cavities. They have reached a 97% trapping rate with single MDA-MB-231 cells spiked in mouse blood [73]. The spiked MDA-MB-231 cells staying in the trapping channel were kept by the continuous blood flow in their delivery channel because their pressure was less than the force is needed that cancer cells can pass through the constriction regions. However, clogging is possible when more CTCs are trapped in the same constriction and it is a drawback to microchannel array.

The CTC-HTECH device is a label-free, low-cost, and size-dependent that enables effectively enrichment of sample in a prostate cancer cell line in blood samples. CTCs enumeration obtained by this method is important because it is a diagnostic or prognostic marker for tumor. The chip can carefully count CTCs in blood and become a trusty tool for clinical settings for rapid CTC capture as well as subsequent counting. The blood cell size distribution, especially WBCs and lymphocytes, in different patients can vary from 8 to 20 μm [74,75]. The differences in deformability between cancer cells and blood cells can be applied in order to WBCs or lymphocytes separation. A new generation of chips is imaginable in which channels with different sizes for constriction regions and the trapping chambers can be fabricated and it may lead to creating a second degree of separation within the same chip. Also, if a more number of CTC exist in the bloodstream, another channels can be added to each row to make it suitable for clinical samples. CTC-HTECH chips can be used in a series mode in identical or different channel dimensions. CTC-HTECH can be connected in a series as multistage multicycle CTC enrichment [62]. CTCs and cfDNA technologies are used parallel in prospective clinical trials, which might be able to explain the current resistance by solid tumors to targeted therapies [48].

9.4 Governing equations

In the following sections, in order to consider the SAW problem, the three general equations of continuity, momentum (Navier-Stokes), and energy in the presence of the SAW are described below. First, considering the fully developed microchannel profiles, the viscose and thermal boundary layer thicknesses are discussed.

The nonlinear relationship of the first-order of velocity, pressure, and temperature fields occurs at a thin thermocouple boundary near the surface. One of the numerical challenges is the dissolving of the thermo viscous boundary layers [76]. The thickness of the thermal boundary layers δ_{th} and the hydrodynamic(viscos) boundary layer δ in a fluid due to the application of an ultrasound wave are as follows:

$$\delta_{th} = \sqrt{\frac{2D}{\omega}} \tag{9.1}$$

$$\delta = \sqrt{\frac{2\vartheta}{\omega}} \tag{9.2}$$

where ω is the angular frequency of the SAW in radians per second and is calculated from the frequency f as follows:

$$\omega = 2\pi f \tag{9.3}$$

where ϑ is a kinematic viscosity and equivalent (μ/ρ), μ and D_{th} are dynamic viscosity and thermal diffusivity of the thermal boundary layer, calculated as follows:

$$D_{th} = \frac{k}{\rho_0 c_p} \tag{9.4}$$

where c_p, ρ_0, and k are heat capacity at constant pressure (j/m^{3}k), equilibrium density (kg/m^3) and thermal conductivity (w/mk).

The thickness of the thermal and viscous boundary layers is usually 1000 times smaller than the microchannel dimensions [76], which requires a fine mesh near the boundaries plus high computational cost. The viscous interaction with this layer drives the acoustic flow in the bulk fluid. The point to note here is that the thickness of the viscous and thermal boundary layer changes with frequency, and as the frequency increases, the boundary layer thickness decreases. It should be noted that the ratio of two thicknesses also yields the Prandtl dimensionless number as follows, which is also important in relation to the thickness of the boundary layer.

$$\frac{\delta}{\delta_{th}} = \sqrt{\frac{\mu c_p}{k}} = \sqrt{\mathrm{Pr}} \tag{9.5}$$

9.4.1 Equations of perturbation theory caused by the acoustic field

Here, the basic perturbation equations are examined in the second order for heat transfer and fluid flow. The first-order equations of fluid flow and heat transfer are not covered, yet they are applied to reach the main governing equations.

In the first-order equations of the acoustic field due to SAWs, the energy equation for temperature (T_1), the kinematic continuity equation for pressure (P_1) and the momentum equation for velocity field (u_1) are described as follows [76]:

$$\frac{\partial T_1}{\partial t} = D_{th} \nabla^2 T_1 + \frac{\alpha_p T_0}{\rho_0 c_p} \frac{\partial p_1}{\partial t} \tag{9.6}$$

$$\frac{\partial p_1}{\partial t} = \frac{1}{\gamma k_s} \left[\alpha_p \frac{\partial T_1}{\partial t} - \nabla u_1 \right] \tag{9.7}$$

$$\rho_0 \frac{\partial u_1}{\partial t} = -\nabla p_1 + \mu \nabla^2 u_1 + \beta \mu \nabla (\nabla u_1) \tag{9.8}$$

where D_{th}, T_0, ρ_0, α_p, γ, k_s, and β are the thermal diffusivity, the wall temperature, mass density, thermal expansion coefficient, specific heat capacity ratio, isentropic compressibility, and the viscosity ratios and are defined as follows:

$$\alpha_p = -\frac{1}{\rho} \left(\frac{\partial \rho}{\partial T} \right)_p \tag{9.9}$$

$$\gamma = \frac{C_p}{C_v} = 1 + \frac{\alpha_p^2 T_0}{\rho_0 C_p k_s} \tag{9.10}$$

$$\beta = \frac{\mu_B}{\mu} + \frac{1}{3} \tag{9.11}$$

where the μ_B and μ are bulk viscosity and dynamic viscosity.

$$k_s = \frac{1}{\rho_0} \left(\frac{\partial \rho}{\partial p} \right)_s \tag{9.12}$$

According to the definition of isentropic sound velocity,

$$c^2 = \left(\frac{\partial p}{\partial \rho}\right)_s = \frac{1}{k_s \rho_0} \rightarrow k_s = \frac{1}{\rho_0 c^2} \tag{9.13}$$

It is assumed that all first-order fields have a harmonic time dependence $e^{-i\omega t}$ due to the ultrasound field. There is a complete derivation of these equations assuming the harmonic time dependence on the source [77] as follows.

Energy equation:

$$i\omega T_1 + D_{th} \nabla^2 T_1 = \frac{\gamma - 1}{\alpha_p} \nabla u_1 \tag{9.14}$$

Momentum equation:

$$i\omega u_1 + v \nabla^2 u_1 + v \left[\beta + i \frac{1}{v\gamma \rho_0 \omega k_s} \right] \nabla(\nabla u_1) = \frac{\alpha_p}{\rho_0 \gamma k_s} \nabla T_1 \tag{9.15}$$

The physical field of pressure and velocity field are obtained by the real part and the imagining part.

9.4.2 Second-order equations

A number of acoustic wave effects were observed in the first-order fields obtained by solving the first-order equations. The first-order theory is inadequate to describe nonlinear behavior of the acoustic field in liquid since the time-averaged first-order fields are zero due to harmonic time dependence. It is necessary to extend the second-order equations of density, pressure, velocity, and temperature as follows:

$$\rho = \rho_0 + \rho_1 + \rho_2 \tag{9.16}$$

$$p = p_0 + p_1 + p_2 \tag{9.17}$$

$$u = u_0 + u_1 + u_2 \tag{9.18}$$

which here symbolizes <x> denotes the time averaging of the function $x\,(t)$ over a complete period T.

$$<x> = \frac{1}{\tau} \int_0^\tau dt\, x(t) \tag{9.19}$$

In contrast, the time averaging of the product of two first term, proportional to cos (ωt) is nonzero because $\langle \cos^2(\omega t) \rangle = \frac{1}{2}$.

By applying the second-order acoustic field in the continuum, momentum, and energy equations and by a series of mathematical operations and time averaging, the second-order equations of a continuum and momentum are described as follows [78]:

$$\partial_t (\rho_0 + \rho_1 + \rho_2) = -\nabla ((\rho_0 + \rho_1 + \rho_2)(v_1 + v_2)) \tag{9.20}$$

$$\partial_t \rho_2 = -\rho_0 \nabla v_2 - \nabla(\rho_1 v_1) \tag{9.21}$$

$$\left(\rho_0 + \rho_1 + \rho_2\right)\partial_t\left(v_1 + v_2\right)$$
$$= -\nabla\left(p_0 + p_1 + p_2\right) - \left(\rho_0 + \rho_1 + \rho_2\right)\left(\left(v_1 + v_2\right)\nabla\right)\left(v_1 + v_2\right) + \eta\nabla^2\left(v_1 + v_2\right) \quad (9.22)$$
$$+ \beta\eta\nabla\left(\nabla\left(v_1 + v_2\right)\right)$$

$$\rho_0\,\partial_t v_2 = -\nabla p_2 + \eta\nabla^2 v_2 + \beta\eta\nabla(\nabla v_2) - \rho_1\,\partial_t v_1 - \rho_0(v_1\nabla)v_1 \quad (9.23)$$

It is obvious that the time averaging of the second-order fields will generally be nonzero, and during the time averaging, the first-order product term will come as the source term on the right of the governing equation.

The nonzero velocity $\langle v_2 \rangle$ is called the acoustic flow, where the bulk fluid moves due to the viscous stresses produced in the fluid near the wall, when the acoustic wave oscillation velocity has to be zero due to the nonslip condition. The nonzero pressure $\langle p_2 \rangle$, increases the acoustic propagation force due to the scattering of the acoustic wave on the particle and causes acoustophoretic motion of the particle.

9.5 Acoustophoretic motion of particles in a PDMS microchannel using SAW

The importance of particle manipulation methods is well known in the last decade. Different methods including passive and active methods were applied so as to improve the manipulation of cells. Precise handling of particles both in active and passive methods depends on the efficient design and correct analyis of fluid-particle interaction and imposed external field in microchannel. SAW as a label-free approach, brings this opportunity to separate biocells efficiently without any damages to cells. Clear surface of PDMS lets exact monitor of cells by microscope and since the acoustic pressure field shape well and efficient in a microchannel with PDMS wall, it made as a popular method to separate or sort biocells.

Numerical investigation of particle motion in a microchannel is studied. In this case, the perpendicular section to fluid flow is considered and acoustofluidic motion is discussed. A rectangular section has 600 μm width and 125 μm height. The working frequency is 6.65 MHz and the wavelength is 600 μm which fits in the channel width. Particle materials are polystyrene and fluid is water. The acoustic contrast factor of polystyrene in the water is positive. Polystyrene diameter is 10 μm with 249 Pa^{-1} compressibility and 1050 kg/m^3 [79]. Water density is 997 kg/m^3 and its compressibility is 448 Pa^{-1} [79]. The section of this channel is illustrated in Fig. 9.10. The effect of the PDMS wall and Lithium Niobate is considered as a boundary condition on the fluid domain.

The primary radiation force imposed on spherical particles to separate cell in the standing ultrasonic field is:

$$F_R = -\left(\frac{\pi P_0^2 V_p \beta_f}{2\lambda}\right)\phi(\beta\rho)\sin\left(\frac{4\pi x}{\lambda}\right) \quad (9.24)$$

$$\phi(\beta\rho) = \frac{5\rho_p - \rho_f}{2\rho_p + \rho_f} - \frac{\beta_p}{\beta_f} \quad (9.25)$$

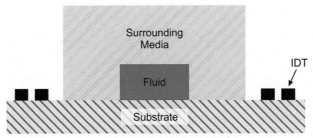

FIGURE 9.10 The schematic of the studied PDMS microchannel.

where P_0, V_p, β_p ρ_p, β_f, and ρ_f are pressure amplitude, particle volume, particle compressibility, particle density, fluid compressibility, and fluid density, respectively. Also λ, x, and ϕ are the wavelength, distance from pressure node and acoustic contrast factor respectively.

When the particles are moving through the microchannel, the surface acoustic field imposed perpendicularly on the fluid flow. The interference of two waves coming from the opposite direction creates a standing wave in the resonance frequency. This standing wave aggregates particles in three regions. Two regions are beside walls and the other is the channel center. Fluid flow by using ARF sort undistributed particles in three regions. Since the particle material is different from fluid, the pressure wave scatters and the gradient of moving energy by the wave will create a force on the particle.

9.5.1 Streamlines

The acoustic streaming line is shown in Fig. 9.11. There are four regions with quasi-symmetrical acoustic streaming.

The strength of acoustic streaming is greater at the bottom of the channel because it is close to the vibrating wall. Since the acoustic volume force is stronger in the acoustic boundary layer, the strength of acoustic streaming is remarkable close to the vibrating wall.

9.5.2 The acoustic streaming V2

The fluid flow (acoustic streaming) after applying an acoustic field is shown in Fig. 9.12. Acoustic waves influence fluid motion and different particles in a fluid.

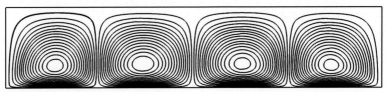

FIGURE 9.11 The acoustic streaming line.

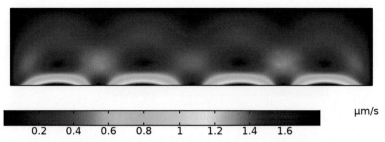

μm/s

FIGURE 9.12 The fluid flow (acoustic streaming) after applying acoustic field.

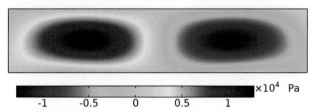

$\times 10^4$ Pa

FIGURE 9.13 The total acoustic pressure.

The impact of acoustic streaming is more on small particles than larger ones. The critical size of the particle which shows ARF or acoustic streaming is dominant depends on the particle and fluid properties. This critical size can have derived after numerical calculation and analysis. Acoustic streaming can be applied to many applications such as micromixing, enhancing heat transfer in microchannel and homogenizing the medicine concentration.

9.5.3 Pressure

Fig. 9.13 depicts the total acoustic pressure. The interference of two SAWs in opposite direction creates standing surface acoustic in the resonance frequency of the device.

The shaped wave is one wavelength in a channel with the exact width of the wavelength. In the three regions including sidewalls and channel center pressure is zero or it is a pressure node while in dark blue and dark red region it is pressure antinode.

9.5.4 Intensity

The ARF depends on the acoustic intensity magnitude. The intensity magnitude, as the main parameter to direct particles, is illustrated in Fig. 9.14. Two regions with high intensity are located in the channel, which is because of the pressure antinode shaped in the channel. Intensity can be evaluated as a parameter to recognize if the device is in the resonance mode. In the resonance mode, the intensity is at a maximum level and the device is working in the optimum condition since the maximum of radiation force is applied.

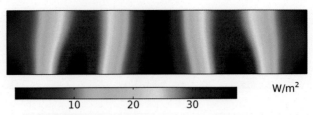

FIGURE 9.14 The acoustic intensity magnitude.

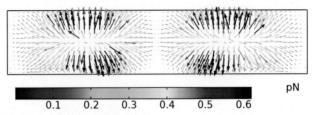

FIGURE 9.15 The size and direction of the force.

9.5.5 Force

Due to the nonlinear behavior of acoustic field, induced pressure is observed. This pressure field will create a force on particles. The size and direction of the force are shown in Fig. 9.15.

The maximum force value is located at a maximum intensity. Because the polystyrene acoustic contrast factor is positive, the radiation force direction for particles is from pressure antinode to node. That's why particles go toward channel wall and center.

9.5.6 Motion of particles

The effect of the acoustic field on the particle motion is depicted in Fig. 9.16.

In $t = 0$ (before applying the acoustic field), particles distributed regularly in the channel. As soon as the acoustic field applied, particles start to move toward pressure nodes. After 8 s they are almost aggregated in pressure nodes. The velocity of particles is under influence of acoustic streaming and ARF. Also the weakness of acoustic streaming comparing to ARF is obvious, the total particle movement pattern is similar to radiation force direction.

Due to nonlinear acoustic behavior in the fluid, the pressure known as acoustic radiation pressure and the velocity known as acoustic streaming is induced. The carrying momentum and energy by the acoustic wave will move particles. The determination of the exact desired particle motion takes experimental and numerical investigation. The radiation force is dominant for larger particles whereas acoustic streaming is dominant for small particles. In working frequency for separation and sorting applications, SAW devices do not have a considerable thermal effect. Acoustic

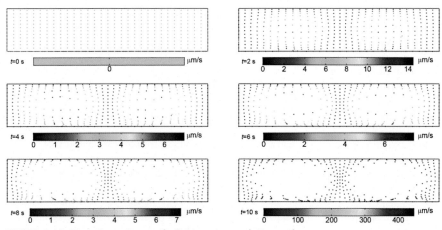

FIGURE 9.16 The effect of acoustic field on the particles motion.

separation is label free and harmless approach which gives scientists this privilege to apply it for all kinds of materials including biocells and viable cancer cells.

References

[1] P. Pop, W.H. Minhass, J. Madsen, Microfluidic Very Large Scale Integration (VLSI), Springer, Cham, Switzerland, (2016).

[2] United States Patent and Trademark Office, search issued patents for "microfluidic" in title or abstract, http://patft.uspto.gov, accessed 2009.

[3] D. Mark, S. Haeberle, G. Roth, F. Von Stetten, R. Zengerle, Microfluidic lab-on-a-chip platforms: requirements, characteristics and applications, Microfluidics Based Microsystems, Springer, Dordrecht, 2010, pp. 305–376.

[4] K. Sritharan, Applications of Surface Acoustic Waves (SAW) for Chemical and Biological Analysis, 2008.

[5] D.B. Go, M.Z. Atashbar, Z. Ramshani, H.C. Chang, Surface acoustic wave devices for chemical sensing and microfluidics: a review and perspective, Anal. Methods 9 (28) (2017) 4112–4134.

[6] Z.V. Guttenberg, A. Rathgeber, S. Keller, J.O. Rädler, A. Wixforth, M. Kostur, et al. Flow profiling of a surface-acoustic-wave nanopump, Phys. Rev. E 70 (5) (2004) 056311.

[7] W. Nyborg, Acoustic Streaming, Academic Press Inc, New York, (1965).

[8] T. Uchida, T. Suzuki, S. Shiokawa, Investigation of acoustic streaming excited by surface acoustic waves, in: 1995 IEEE Ultrasonics Symposium. Proceedings. An International Symposium, IEEE, November 1995, vol. 2, pp. 1081–1084.

[9] A. Wixforth, Acoustically driven programmable microfluidics for biological and chemical applications, JALA-J. Assoc. Lab. Autom. 11 (6) (2006) 399–405.

[10] S.C.S. Lin, X. Mao, T.J. Huang, Surface acoustic wave (SAW) acoustophoresis: now and beyond, Lab Chip 12 (16) (2012) 2766–2770.

[11] X. Ding, P. Li, S.C.S. Lin, Z.S. Stratton, N. Nama, F. Guo, et al. Surface acoustic wave microfluidics, Lab Chip 13 (18) (2013) 3626–3649.

[12] G. Destgeer, H.J. Sung, Recent advances in microfluidic actuation and micro-object manipulation via surface acoustic waves, Lab Chip 15 (13) (2015) 2722–2738.

[13] B. Jakoby, M.J. Vellekoop, Properties of love waves: applications in sensors, Smart Mater. Struct. 6 (6) (1997) 668.

[14] O. Tigli, M.E. Zaghloul, Temperature stability analysis of CMOS-saw devices by embedded heater design, IEEE Trans. Dev. Mater. Reliab. 8 (4) (2008) 705–713.

[15] S. Anand, J. Nylk, S.L. Neale, C. Dodds, S. Grant, M.H. Ismail, et al. Aerosol droplet optical trap loading using surface acoustic wave nebulization, Opt. Express 21 (25) (2013) 30148–30155.

[16] A. Winkler, R. Brünig, C. Faust, R. Weser, H. Schmidt, Towards efficient surface acoustic wave (SAW)-based microfluidic actuators, Sens. Actuators A Phys. 247 (2016) 259–268.

[17] P.S. Dittrich, A. Manz, Lab-on-a-chip: microfluidics in drug discovery, Nat. Rev. Drug Discov. 5 (3) (2006) 210.

[18] R. Ma, L. Xie, C. Han, K. Su, T. Qiu, L. Wang, et al. In vitro fertilization on a single-oocyte positioning system integrated with motile sperm selection and early embryo development, Anal. Chem. 83 (8) (2011) 2964–2970.

[19] J. Lacombe, S.L. Phillips, F. Zenhausern, Microfluidics as a new tool in radiation biology, Cancer Lett. 371 (2) (2016) 292–300.

[20] K.W. Pulsipher, D.A. Hammer, D. Lee, C.M. Sehgal, Engineering theranostic microbubbles using microfluidics for ultrasound imaging and therapy: a review, Ultrasound Med. Biol. 44 (12) (2018) 2441–2460.

[21] A. Boussommier-Calleja, R. Li, M.B. Chen, S.C. Wong, R.D. Kamm, Microfluidics: a new tool for modeling cancer–immune interactions, Trends Cancer 2 (1) (2016) 6–19.

[22] A. Barg, R. Ossig, T. Goerge, M.F. Schneider, H. Schillers, H. Oberleithner, et al. Soluble plasma-derived von Willebrand factor assembles to a haemostatically active filamentous network, Thromb. Haemost. 97 (04) (2007) 514–526.

[23] B. Kasemo, Biological surface science, Surf. Sci. 500 (1–3) (2002) 656–677.

[24] N. Li, A. Tourovskaia, A. Folch, Biology on a chip: microfabrication for studying the behavior of cultured cells, Crit. Rev. Biomed. Eng. 31 (5&6) (2003) 423–488.

[25] S.K. Sia, G.M. Whitesides, Microfluidic devices fabricated in poly (dimethylsiloxane) for biological studies, Electrophoresis 24 (21) (2003) 3563–3576.

[26] D.J. Beebe, G.A. Mensing, G.M. Walker, Physics and applications of microfluidics in biology, Ann. Rev. Biomed. Eng. 4 (1) (2002) 261–286.

[27] D.R. Reyes, D. Iossifidis, P.A. Auroux, A. Manz, Micro total analysis systems. 1. Introduction, theory, and technology, Anal. Chem. 74 (12) (2002) 2623–2636.

[28] N.T. Nguyen, Z. Wu, Micromixers—a review, J. Micromech. Microeng. 15 (2) (2005) R1.

[29] T.M. Squires, S.R. Quake, Microfluidics: fluid physics at the nanoliter scale, Rev. Mod. Phys. 77 (3) (2004) 977.

[30] S.J.J. Lee, N. Sundararajan, Microfabrication for Microfluidics, Artech House, (2010).

[31] S. Götz, U. Karst, Recent developments in optical detection methods for microchip separations, Anal. Bioanal. Chem. 387 (1) (2007) 183–192.

[32] G. Khanarian, Optical properties of cyclic olefin copolymers, Opt. Eng. 40 (6) (2001) 1024–1030.

[33] P. Laval, N. Lisai, J.B. Salmon, M. Joanicot, A microfluidic device based on droplet storage for screening solubility diagrams, Lab Chip 7 (7) (2007) 829–834.

[34] J.M. Bartlett, D. Stirling, A short history of the polymerase chain reaction, PCR Protocols, Humana Press, Totowa, New Jersey, 2003, pp. 3–6.

[35] V.N. Goral, C. Zhou, F. Lai, P.K. Yuen, A continuous perfusion microplate for cell culture, Lab Chip 13 (6) (2013) 1039–1043.

[36] X. Ding, P. Li, S.C.S. Lin, Z.S. Stratton, N. Nama, F. Guo, et al. Surface acoustic wave microfluidics, Lab Chip 13 (18) (2013) 3626–3649.

[37] H.C. Sambrook, Molecular Cloning: A Laboratory Manual, Cold Spring Harbor, NY, (1989).

[38] S.A. Peyman, A. Iles, N. Pamme, Rapid on-chip multi-step (bio) chemical procedures in continuous flow–manoeuvring particles through co-laminar reagent streams, Chem. Commun. (10) (2008) 1220–1222.

[39] D. Carugo, T. Octon, W. Messaoudi, A.L. Fisher, M. Carboni, N.R. Harris, et al. A thin-reflector microfluidic resonator for continuous-flow concentration of microorganisms: a new approach to water quality analysis using acoustofluidics, Lab Chip 14 (19) (2014) 3830–3842.

[40] J.J. Hawkes, R.W. Barber, D.R. Emerson, W.T. Coakley, Continuous cell washing and mixing driven by an ultrasound standing wave within a microfluidic channel, Lab Chip 4 (5) (2004) 446–452.

[41] X. Ding, S.C.S. Lin, B. Kiraly, H. Yue, S. Li, I.K. Chiang, et al. On-chip manipulation of single microparticles, cells, and organisms using surface acoustic waves, Proc. Natl. Acad. Sci. 109 (28) (2012) 11105–11109.

[42] Y. Chen, X. Ding, S.C. Steven Lin, S. Yang, P.H. Huang, N. Nama, et al. Tunable nanowire patterning using standing surface acoustic waves, ACS Nano 7 (4) (2013) 3306–3314.

[43] X. Ding, Z. Peng, S.C.S. Lin, M. Geri, S. Li, P. Li, et al. Cell separation using tilted-angle standing surface acoustic waves, Proc. Natl. Acad. Sci. 111 (36) (2014) 12992–12997.

[44] J. Nam, H. Lim, C. Kim, J. Yoon Kang, S. Shin, Density-dependent separation of encapsulated cells in a microfluidic channel by using a standing surface acoustic wave, Biomicrofluidics 6 (2) (2012) 024120.

[45] J. Nam, H. Lim, D. Kim, S. Shin, Separation of platelets from whole blood using standing surface acoustic waves in a microchannel, Lab Chip 11 (19) (2011) 3361–3364.

[46] X. Bossuyt, G.E. Marti, T.A. Fleisher, Comparative analysis of whole blood lysis methods for flow cytometry, Cytom. J. Int. Soc. Anal. Cytol. 30 (3) (1997) 124–133.

[47] S. Chow, D. Hedley, P. Grom, R. Magari, J.W. Jacobberger, T.V. Shankey, Whole blood fixation and permeabilization protocol with red blood cell lysis for flow cytometry of intracellular phosphorylated epitopes in leukocyte subpopulations, Cytom. A-J. Int. Soc. Anal. Cytol. 67 (1) (2005) 4–17.

[48] T. Luo, L. Fan, R. Zhu, D. Sun, Microfluidic single-cell manipulation and analysis: methods and applications, Micromachines 10 (2) (2019) 104.

[49] J. Wyckoff, B. Gligorijevic, D. Entenberg, J. Segall, J. Condeelis, High-resolution multiphoton imaging of tumors in vivo, Cold Spring Harbor Protocols 2011 (10) (2011) pdb-top065904.

[50] E. Sahai, Illuminating the metastatic process, Nat. Rev. Cancer 7 (10) (2007) 737.

[51] A.F. Chambers, A.C. Groom, I.C. MacDonald, Metastasis: dissemination and growth of cancer cells in metastatic sites, Nat. Rev. Cancer 2 (8) (2002) 563.

[52] C.T. Kuo, H.K. Liu, G.S. Huang, C.H. Chang, C.L. Chen, K.C. Chen, et al. A spatiotemporally defined in vitro microenvironment for controllable signal delivery and drug screening, Analyst 139 (19) (2014) 4846–4854.

[53] J.W. Song, S.P. Cavnar, A.C. Walker, K.E. Luker, M. Gupta, Y.C. Tung, et al. Microfluidic endothelium for studying the intravascular adhesion of metastatic breast cancer cells, PLoS ONE 4 (6) (2009) e5756.

[54] J.S. Jeon, I.K. Zervantonakis, S. Chung, R.D. Kamm, J.L. Charest, In vitro model of tumor cell extravasation, PLoS ONE 8 (2) (2013) e56910.

[55] R. Riahi, Y.L. Yang, H. Kim, L. Jiang, P.K. Wong, Y. Zohar, A microfluidic model for organ-specific extravasation of circulating tumor cells, Biomicrofluidics 8 (2) (2014) 024103.

[56] I.J. Fidler, The pathogenesis of cancer metastasis: the'seed and soil'hypothesis revisited, Nat. Rev. Cancer 3 (6) (2003) 453.

[57] K.H. Benam, S. Dauth, B. Hassell, A. Herland, A. Jain, K.J. Jang, et al. Engineered in vitro disease models, Annu. Rev. Pathol. 10 (2015) 195–262.

[58] S. Turcotte, S.A. Rosenberg, Immunotherapy for metastatic solid cancers, Adv. Surg. 45 (1) (2011) 341–360.

[59] S. Kannan, N. Venugopal, Current trends in microfluidics for single cell isolation in cancer diagnostics enabling downstream proteomics applications, MOJ Proteomics Bioinform. 3 (4) (2016) 98-L106.

[60] M.J. Markuszewski, R. Kaliszan, Using bioanalysis for cancer diagnosis and prognosis, Bioanalysis 6 (7) (2014) 907–909.

[61] C. Bayarri-Lara, F.G. Ortega, A.C.L. de Guevara, J.L. Puche, J.R. Zafra, D. de Miguel-Pérez, et al. Circulating tumor cells identify early recurrence in patients with non-small cell lung cancer undergoing radical resection, PLoS ONE 11 (2) (2016) e0148659.

[62] X. Ren, B.M. Foster, P. Ghassemi, J.S. Strobl, B.A. Kerr, M. Agah, Entrapment of prostate cancer circulating tumor cells with a sequential size-based microfluidic chip, Anal. Chem. 90 (12) (2018) 7526–7534.

[63] F.F. Becker, X.B. Wang, Y. Huang, R. Pethig, J. Vykoukal, P.R. Gascoyne, Separation of human breast cancer cells from blood by differential dielectric affinity, Proc. Natl. Acad. Sci. 92 (3) (1995) 860–864.

[64] P.R. Gascoyne, J. Noshari, T.J. Anderson, F.F. Becker, Isolation of rare cells from cell mixtures by dielectrophoresis, Electrophoresis 30 (8) (2009) 1388–1398.

[65] P. Augustsson, C. Magnusson, M. Nordin, H. Lilja, T. Laurell, Microfluidic, label-free enrichment of prostate cancer cells in blood based on acoustophoresis, Anal. Chem. 84 (18) (2012) 7954–7962.

[66] A.H. Yang, H.T. Soh, Acoustophoretic sorting of viable mammalian cells in a microfluidic device, Anal. Chem. 84 (24) (2012) 10756–10762.

[67] S. Shim, K. Stemke-Hale, J. Noshari, F.F. Becker, P.R. Gascoyne, Dielectrophoresis has broad applicability to marker-free isolation of tumor cells from blood by microfluidic systems, Biomicrofluidics 7 (1) (2013) 011808.

[68] S. Shim, K. Stemke-Hale, J. Noshari, F.F. Becker, P.R. Gascoyne, Dielectrophoresis has broad applicability to marker-free isolation of tumor cells from blood by microfluidic systems, Biomicrofluidics 7 (1) (2013) 011807.

[69] A.F. Sarioglu, N. Aceto, N. Kojic, M.C. Donaldson, M. Zeinali, B. Hamza, et al. A microfluidic device for label-free, physical capture of circulating tumor cell clusters, Nat. Methods 12 (7) (2015) 685.

[70] H. Mohamed, L.D. McCurdy, D.H. Szarowski, S. Duva, J.N. Turner, M. Caggana, Development of a rare cell fractionation device: application for cancer detection, IEEE Trans. Nanobiosci. 3 (4) (2004) 251–256.

[71] S. Zheng, H. Lin, J.Q. Liu, M. Balic, R. Datar, R.J. Cote, et al. Membrane microfilter device for selective capture, electrolysis and genomic analysis of human circulating tumor cells, J. Chromatogr. A 1162 (2) (2007) 154–161.

[72] G. Vona, A. Sabile, M. Louha, V. Sitruk, S. Romana, K. Schütze, et al. Isolation by size of epithelial tumor cells: a new method for the immunomorphological and molecular characterization of circulating tumor cells, Am. J. Pathol. 156 (1) (2000) 57–63.

[73] Y. Yoon, J. Lee, K.C. Yoo, O. Sul, S.J. Lee, S.B. Lee, Deterministic capture of individual circulating tumor cells using a flow-restricted microfluidic trap array, Micromachines 9 (3) (2018) 106.

[74] W.J. Allard, J. Matera, M.C. Miller, M. Repollet, M.C. Connelly, C. Rao, et al. Tumor cells circulate in the peripheral blood of all major carcinomas but not in healthy subjects or patients with nonmalignant diseases, Clin. Cancer Res. 10 (20) (2004) 6897–6904.

[75] F. Krombach, S. Münzing, A.M. Allmeling, J.T. Gerlach, J. Behr, M. Dörger, Cell size of alveolar macrophages: an interspecies comparison, Environ. Health Perspect. 105 (Suppl. 5) (1997) 1261–1263.

[76] P.B. Muller, R. Barnkob, M.J.H. Jensen, H. Bruus, A numerical study of microparticle acoustophoresis driven by acoustic radiation forces and streaming-induced drag forces, Lab Chip 12 (22) (2012) 4617–4627.

[77] M. Gedge, M. Hill, Acoustofluidics 17: theory and applications of surface acoustic wave devices for particle manipulation, Lab Chip 12 (17) (2012) 2998–3007.

[78] M. Ghassemi, A. Shahidian, Nano and Bio Heat Transfer and Fluid Flow, Academic Press, London, (2017).

[79] P.B. Muller, M. Rossi, A.G. Marin, R. Barnkob, P. Augustsson, T. Laurell, et al. Ultrasound-induced acoustophoretic motion of microparticles in three dimensions, Phys. Rev. E 88 (2) (2013) 023006.

Index

Note: Page numbers followed by "f" indicate figures, "t" indicate tables.

Printed in the United States
By Bookmasters